The Peasant Cotton Revolution in West Africa
Côte d'Ivoire, 1880–1995

The literature on Africa is dominated by accounts of crisis, doom and gloom, but this book presents one of the few long-running success stories. Thomas Bassett, a distinguished American geographer well known in the field of development, tells an unusual story of the growth of the cotton economy of West Africa, where change was brought about by tens of thousands of small-scale peasants farmers. While the introduction of new strains of cotton in francophone West Africa was in part the result of agronomic research by French scientists, supported by an unusually efficient marketing structure, this is not a case of triumphant top-down "planification." Employing the case of Côte d'Ivoire, Professor Bassett shows agricultural intensification to result from the cumulative effect of decades of incremental changes in farming techniques and social organization. A significant contribution to the literature, the book demonstrates the need to consider the local and temporal dimensions of agricultural innovations. It brings into question many key assumptions that have influenced development policies during the twentieth century.

THOMAS J. BASSETT is Associate Professor of Geography at the University of Illinois, Urbana-Champaign. He is co-author of *Land in African Agrarian Systems* (1993) and *Maps of Africa to 1900* (2000), and has been engaged in long-term field work in Côte d'Ivoire since 1981.

African Studies Series 101

Published in collaboration with
THE AFRICAN STUDIES CENTRE, CAMBRIDGE

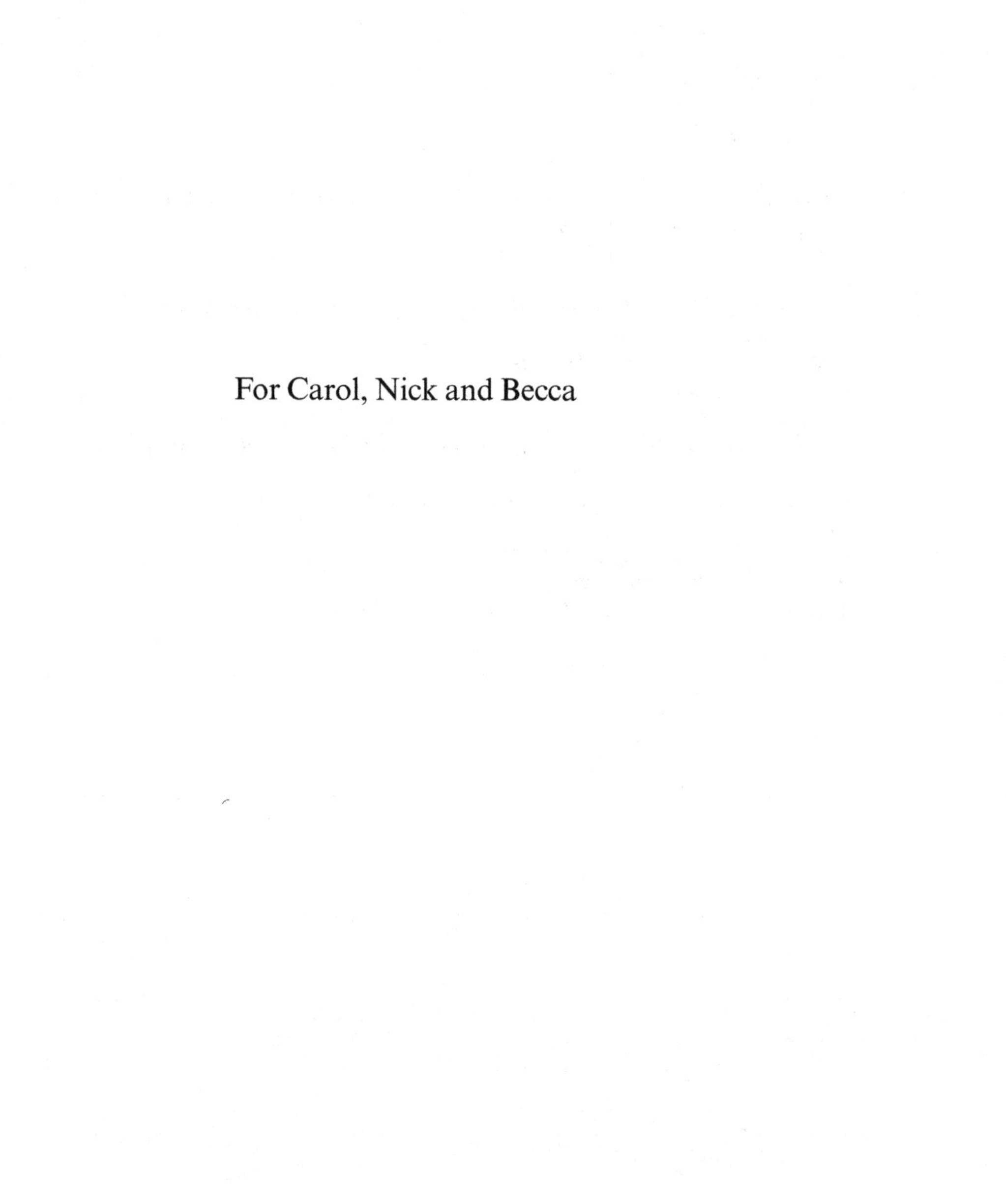

For Carol, Nick and Becca

The Peasant Cotton Revolution in West Africa

Côte d'Ivoire, 1880–1995

Thomas J. Bassett

CAMBRIDGE UNIVERSITY PRESS
Cambridge, New York, Melbourne, Madrid, Cape Town, Singapore, São Paulo

Cambridge University Press
The Edinburgh Building, Cambridge CB2 2RU, UK

Published in the United States of America by Cambridge University Press, New York

www.cambridge.org
Information on this title: www.cambridge.org/9780521783132

First published 2001
This digitally printed first paperback version 2006

A catalogue record for this publication is available from the British Library

ISBN-13 978-0-521-78313-2 hardback
ISBN-10 0-521-78313-5 hardback

ISBN-13 978-0-521-78883-0 paperback
ISBN-10 0-521-78883-8 paperback

Contents

Illustrations

Plates

Figures

Tables

Preface

Over the years in which I have lived and conducted research in the community of Katiali, people have frequently asked me: "What are you going to do with this information?" From the start, most people were wary of my presence in their village. Just forty years earlier, when colonial administrators asked questions about household size and production, the information they collected was used to determine the number of laborers and sacks of grain the village was forced to provide for the "development" (*la mise en valeur*) of the colony. Many people still held vivid memories of the suffering and injustices experienced during this time.

I responded to these questions by saying that my country "gave" millions of dollars each year to countries like Côte d'Ivoire to help finance development projects. The problem was that too often decisions on how to allocate funds were made by people who had little idea of what conditions were like in places like Katiali. I was there, asking questions, so that I could go back to my country and influence the decision makers.

In retrospect, my response was earnest but ultimately naive. First, it assumed that aid donors would in fact listen to what the farmers of Katiali had to tell me about their farming systems, and the obstacles they faced to improving their livelihoods. I soon learned that it was not going to be that easy. At the World Bank's regional headquarters in Abidjan, it was difficult just to get past the reception desk. I felt like a trespasser in a gated professional community; muddy-booted researchers without connections were rarely permitted entry. When a sympathetic soul did open his or her door, I saw that most of the documents on the desk before me had the words "Confidential: For Official Use Only" stamped across the front of their gray covers. In short, both the guarded entry and the conditions of secrecy hindered my efforts to get a dialogue going on rural development issues pertinent to the people of Katiali. Although the bank has opened up more during the past decade, its structural adjustment policies still show an apparent deafness to the voices of peasant farmers.

I was also naive about the capacity of individuals and groups like the farmers of Katiali to influence agricultural policies by their own actions.

The literature on African agriculture in the 1980s suggested just the opposite; its dominant themes of agricultural stagnation and disillusionment did not put peasant agency in a favorable light. Peasants were portrayed in largely passive terms, as either retreating from state-controlled markets or falling victim to the exploitation of merchants and industrial capital. The structures and logic of capital accumulation seemed to determine the course of events. Peasant agency amounted to little more than passive resistance to the exactions made on them by a monolithic state and metropolitan interests. At best, farmers were presented as given to experimenting with new crops based on their adaptiveness to local agro-ecological and social conditions, notably labor supplies. Yet, when it came to adopting innovations that would radically alter finely calibrated farming systems, these same farmers were shown to be conservative and unwilling to take risks. These portraits, I found, failed to capture the diverse ways in which peasants have actively participated in the making and remaking of their agricultural systems, to the point that such systems show little resemblance to how they appeared a generation earlier. Contrary to the image of agricultural stagnation, peasant farmer engagement with cotton in West Africa suggests that agricultural change has been momentous. The main argument of this book is that the dynamics of agrarian change have been driven by repeated contestation, negotiation, and innovation among farmers and a host of cotton developers (administrators, agricultural experts, cotton companies, merchants, and aid donors), as well as between them.

The book's title seeks to draw attention to peasant farmer influences in shaping a cotton revolution in West Africa. My primary goal is to present an alternative agricultural and social history to the dominant development narrative, which portrays peasants as the simple recipients of technological innovations conceived and diffused by Western development experts. This heroic view of agrarian change informs the cotton development discourse on West Africa. Despite the populist veneer to this book's title, this story has not been written as a simple counter-narrative that emphasizes peasant ingenuity in resisting or reworking state and private initiatives to intensify cotton. Rather, I attempt to bring out the multiplicity of interests and actors implicated in this story, and the changing circumstances and strategies that have contributed to its unfolding. Farmers have not simply reacted to external interventions; they have played important roles in creating the very conditions (farming systems, policies, and institutions) through which the cotton revolution has emerged. In this sense, peasant farmers of northern Côte d'Ivoire have been central to the making of their own agrarian history.

In doing the research for this book, I have been fortunate to have enjoyed the moral, intellectual, and financial support of many friends, colleagues, and institutions. My deepest gratitude is to my wife, Carol Spindel, for joining me on the many research trips to Côte d'Ivoire and for helping me through critical periods of my fieldwork. Her assistance and companionship were vital to the successful completion of this project. Her own book, *In the Shadow of the Sacred Grove*, offers insights into the people and community of Katiali (Kalikaha), which complement this book in many interesting ways. I also want to thank our children, Nicholas and Rebecca, who, as toddlers and pre-teens, integrated themselves amazingly well into village life during our extended stays.

Michael Watts, my Ph.D. advisor at the University of California, Berkeley, was and continues to be an inspiration and guide through the muddy terrain of development geography, agrarian politics, and agricultural change in Africa. His unflagging moral and intellectual support is deeply appreciated. I must also thank John Sutter and Marie-Hélène Collion, who freely shared their friendship and field experiences with me at the very formative stages of this project.

For financial assistance, I am very grateful to the Social Science Research Council and the American Council of Learned Societies for awarding me an International Doctoral Research Fellowship which allowed me to live in Africa for eighteen months in 1981–82 and to write up a large part of my dissertation. I owe thanks to the Regents of the University of California for a Traveling Research Fellowship in 1981, to International Programs and Studies at the University of Illinois for awarding me a William and Flora Hewlett Summer International Research Grant for summer research in 1986 and 1991, and to the University of Illinois Research Board for field research grants in 1985, 1987, and 1989. I am also thankful to the National Geographic Society, which offered generous support in the summer of 1988; to the Fulbright Senior Research Scholar Program for a research fellowship in the spring and summer of 1992; and to the John D. and Catherine T. MacArthur Foundation for a Collaborative Studies Grant for the years 1993–98. Finally, this longitudinal research project could not have been accomplished without the collective support of my parents, Robert I. and Margaret M. Bassett, and my in-laws, Christine Spindel and the late Murray A. Spindel.

In Côte d'Ivoire, I was fortunate to be affiliated with the Centre Ivoirien de Recherches Economiques et Sociales and the Institut de Géographie Tropicale at the Université de Cocody. I owe special thanks to Achi Atsain, Thomas Eponou, Yves Leon, Jacques Pegatienan-Hiey, and Joseph Yao, for both facilitating my stay and orienting me in the country. My colleagues at the Université de Cocody have been most welcoming and supportive in this effort. Thanks to Simon-Pierre Akanza, Sinaly Coulibaly, Mameri Camara, Atta Koffi, Tiékoura Koné, Tiona Ouattara, and Koli Bi Zuéli.

Other Ivorian researchers with whom I have collaborated at different times include Zoumana Coulibaly, Papa Samba N'Daw, Soro Nambegué, and Kolo Yeo. I give special thanks to the late Philip Ravenhill and Albert Votaw, and to Judy Timyan, Esti Votaw, Mariam Touré, and David and Laura Hess for their hospitality and friendship in Abidjan. At the World Bank's regional office in Abidjan, I want to thank Jean-Paul Chausse, John MacIntire, and Vijaya MacKrandilal for opening their doors and sharing information with me. Thanks, too, to Bruce Zanin and Antoine Anzele of the Foreign Agricultural Service of the United States Department of Agriculture for their excellent on-line reports and e-mail messages on the fast moving developments in the Ivorian cotton sector during the 1990s. I am also thankful to a number of European development professionals who shared their knowledge and opened their homes to me and my family, including Alain Escafre, Jean-Baptiste and Claude Bonnet, Albert Kientz, Xavier Le Roy, Yves Bigot, Patrick Bisson, Jean César, Siegfried Schröeder-Breitschuh and Ulrike Breitschuh. I also wish to acknowledge Mr. and Mrs. Banga Koné for allowing me to stay in their Korhogo home during my weekend visits and for introducing me to their family and friends, especially the late Zoumana Koné. Mrs. Diomandé Ramatou, the archivist at the Compagnie Ivoirienne pour le Développement des Textiles (CIDT) in Bouaké, and Mr. Kouadio Kouadio, CIDT's regional director for the Korhogo area, generously shared their time and documentation with me.

In France I was welcomed by the Centre d'Etudes Africaines in Paris for the 1995–96 academic year, when this book finally began to take shape. I am grateful to Roger Botte, Jean Boutrais and Jean Schmitz for the invitation and office space, and to Chantal Blanc-Pammard and Jean-Pierre Dozon for their warm welcome and assistance throughout the year. Jean Boutrais's reading of the entire manuscript significantly contributed to the revision process. I am especially honored to have this work appear in French in the *à travers champs* series edited by Jean Boutrais and published by the Institut de Recherche pour le Développement (ex-ORSTOM). At the Compagnie Française pour le Développement de Fibres Textiles (CFDT), I greatly profited from discussions with François Béroud and François Geay, whose comparative knowledge of cotton growing in West Africa helped me place the Ivorian case study within a wider regional context. Sylvie Retana graciously responded to my requests for documentation from CFDT. Claude Meillassoux has been both a mentor and a dear friend, whose questions and reflections have made a lasting impression on me and this book. Pierrette and Albert Kohen's friendship and hospitality have made my trips to Paris seem like going home.

For mapping the way to the archives in Africa and France, special thanks to Tim Weiskel, and especially Richard Roberts. Richard's own work on the social history of cotton in Africa, and our ongoing comparative discussions, have shaped this book in significant ways. He also read the complete manuscript and offered valuable suggestions for its subsequent revision. Discussions with Sara Berry, Bonnie Campbell, Jane Guyer, and Alan Isaacman on the political economy and social history of cotton in Africa have also been important to my thinking on these issues. I presented an earlier version of chapter three at a symposium organized by Alan Isaacman and Richard Roberts at the University of Minnesota in May 1992; this later appeared in their edited collection *Cotton, Colonialism, and Social History* published by Heinemann in 1995. Thanks to Soumailia Jakité for teaching me the basics of Bamana/Jula, the *lingua franca* of Côte d'Ivoire. My brother, Robert L. Bassett, deserves special acknowledgment for his life-long encouragement.

At the University of Illinois at Urbana-Champaign, I have enjoyed many opportunities to present this research to colleagues and students whose critical comments and suggestions have contributed to this book's current form. Thanks especially to Alex Winter-Nelson and Leslie Gray, who read the entire manuscript, and to Mahir Saul, with whom I have discussed many of the ideas. My discussions with Rachel Schurmann, Michael Goldman, and William Munroe have influenced my thinking at various junctures. The Center for African Studies has given me continuous support under the leadership of Charles Stewart, Donald Crummey, and Paul Tiyambe Zeleza. The Department of Geography has been exceptional in creating a work environment that blends teaching and research in productive and satisfying ways. I must give a special thanks to the staff cartographer, Jane Domier, who produced many of the maps illustrating this text, and to Barbara Cohen who provided the index.

If it were not for Adama Koné, my research assistant of the past fifteen years, this book would never have been written. Adama's patience, stamina, and good humor made him the perfect diplomat for negotiating interviews in Katiali. I am most indebted to our wonderful "*jatigi*," Donisongui Silué and his family, for hosting us in the village all these years. My understanding of the functioning of village-level farmer organizations has been greatly facilitated by Zéa Silué and Karim Koné, the secretaries of the GVC and COOPAG-CI cooperatives. It was because of the kind welcome of Katienen'golo Silué and the late Bêh Tuo that I chose Katiali as my research site. They were also key informants for many historical details related to agricultural and cultural change. Finally, I thank the residents of Katiali. I hope this book succeeds in highlighting your unsung contributions to the making of an agricultural revolution in Africa.

To all, thank you. May all that you have given to me be returned to you a hundredfold.

Glossary

foroba foro	a collective field worked by members of the same descent group or *kabila* among the Jula
jamu	a family name identifying a Jula individual as belonging to a certain clan or patronymic group (e.g. Koné, Diabaté)
jasa	during the colonial period, a residential area located in the central administrative center (e.g. Korhogo) where the representatives of canton chiefs resided
jongarri	a personal field worked by a Jula individual
kabila	a Jula residential quarter
kabilatigi	the head of a Jula residential quarter
kagon	a personal field worked by a Senufo individual
kagonbile	a collective field headed by a Senufo residential quarter chief
karamoko	a learned Islamic scholar among the Jula
katienetio	a wife given by the parents or maternal uncle of the woman to a Senufo man "for doing a good deed"
katiolo	a Senufo residential quarter
katiolofolo	a Senufo residential quarter chief
kékourougou	a marriage arrangement, formerly common among the Nafara subgroup of the Senufo, in which the wife remains attached to her relatives' production unit and often lives in a separate village from her husband
korofonwala	a period in the Senufo agricultural calendar meaning "everyone is hungry"
koulon pigué	a house captive among the Senufo; usually the descendant of a trade or war captive
lon	the Jula secret initiation society
londen	Jula initiates
lu	a subdivision of the Jula *kabila* that is also the basic unit of production
lutigi	the head of a Jula *lu*

nerbatio	a wife given by the head of a Senufo matrilineage to a man of the lineage
n'golon	a system of reciprocal labor exchanges among the Senufo
poro	the age-class secret initiation society of the Senufo
sando	a Senufo diviner
segbo	the Kiembara Senufo term for a collective field
segnon	a lineage or multi-lineage collective field among the Kasambélé Senufo
segnontio	a wife given by the Senufo village chief to a man for working in the collective field
sinzanga	a sacred grove of the Senufo
sinzangafolo	the spiritual head of a Senufo sacred grove
so	a group of contiguous *lu* among the Jula
tarfolo	the Senufo land priest and guardian of lineage land
tofotio	a wife given by a parent or maternal uncle to a Senufo man
ton	a collective work group among the Jula
tugubélé	spirits of the bush, generally ambivalent towards humans
tyolo	an initiate of the Senufo *poro* society during the final six-year cycle
woroso	a house captive among the Jula who was born into a slave family

1 Introduction

> The key to the relative success of cotton development in francophone Africa has been the ability of the industry to maintain effective coordination among the different layers of participants . . . By and large, the French CFDT, through its various interventions – ranging from upstream research to downstream marketing assistance – has helped alleviate most constraints and risks to the cotton sectors of its recipient countries.
>
> U. Lele, *et al.*, 1989

> The texts of development have always been avowedly strategic and tactical – promoting, licensing and justifying certain interventions and practices, delegitimizing and excluding others . . . what do the texts of development not say? What do they suppress? Who do they silence – and why?
>
> J. Crush, 1995

Cotton is hailed as one of the rare agricultural success stories in contemporary Africa.[1] In what is conventionally portrayed as a bleak agricultural landscape during the 1970s and 1980s,[2] the intensification of cotton, particularly in francophone West Africa, has been proclaimed as nothing less than "spectacular."[3] Like most development narratives, this story has a beginning, middle, and end.[4] The core narrative begins like this. In the early 1960s poor and illiterate peasant farmers intercropped low-yielding varieties of cotton with food crops. It was at this time that French agronomists working in experiment stations introduced a technological package comprised of a high-yielding variety, pesticides, and fertilizers that promised to catapult cotton growing to undreamed of production levels. In concert with the newly independent state, La Compagnie française pour le développement des fibres textiles (CFDT) (the French Company for the Development of Textile Fibers) established a new institutional apparatus that proved to be highly effective in transferring the new technology to an increasingly large number of cotton growers. The vertically integrated research-production-marketing structure known as "the CFDT system" supplied inputs, purchased cotton, and paid producers in a timely manner, which contrasted with the relatively poorer performance of anglophone Africa.[5] At the farm level, this growth has been linked to increasing yields per hectare, an increasing area cultivated per grower, and an ever expanding number of growers. In contrast to the colonial period, when yields were low and farmers hostile to cotton development initiatives, the independence period has witnessed a revolution in cotton growing involving the participation of thousands of peasant farmers.

There are recurring themes as well as silences running through this dominant narrative. The salient themes center around technological and institutional innovations, incentives to growers, and the importance of foreign expertise in constructing this revolution. The CFDT-induced and technology-led model of cotton development is trumpeted as the principal player behind the impressive growth of cotton during the 1970s and 1980s. The most deafening silences surround the roles played by peasant farmers. They are, by and large, presented as a passive and muted lot that simply responds to outside stimuli by changing their farming practices. Teleology replaces history to account for the agricultural and social transformations at the heart of this agricultural development story.

The central argument of this book is that the cotton revolution in Côte d'Ivoire has been shaped in fundamental ways by African peasant farmers as they struggled with a variety of agents (e.g. colonial administrators, merchants) as well as among themselves over the direction of these changes. In contrast to conventional depictions of peasants either as dupes who are somehow enticed into participating in cotton development programs,[6] or as passive bearers of a mode of production who react in predictable ways to such pressures as adverse terms of trade,[7] this book presents a more dynamic portrait of peasants, whose past and present actions provide a framework within which developers must work. From a social theoretical perspective, it is untenable to view peasant farmers independently of the institutional structures (e.g. "the CFDT system") of which they are a part. Peasants have not simply responded to the incentives and technological packages introduced by the cotton company and the Ivorian government. To the contrary, they have had, and continue to have, a hand in shaping the form and content of these institutions, packages, and shifting policies through their actions.

The vagaries of cotton policies and production also reflected conflicts among colonial administrators, agricultural development experts, merchants, and textile company representatives over how to intensify cotton. Disagreements were common among administrators over how to motivate peasants to produce cotton for export markets. Some believed force was necessary, others argued that competitive prices would be sufficient. While some officials doggedly pushed cotton, others promoted yams and rice.

Some of the resulting contradictions were stunning. The coexistence of forced cultivation and "free trade" allowed peasants to sell their crop at competitive prices to local Jula merchants, who supplied cotton fiber to the local weaving industry. Colonial administrators were often critical of European merchants who refused to offer higher prices. At the same time, agricultural experts disagreed with the administrative decision to grow cotton as a sole crop in large fields, arguing that it performed better as an intercrop in food crop fields.

Disagreements and tensions within rural communities were also characteristic of the process of agricultural intensification during the colonial and post-colonial periods. Conflicts over the control of labor and cotton profits erupted, especially between juniors and seniors and between men and women. The break up of large production units, the erosion of the *poro* initiation society among the Senufo, and the diffusion of ox-plows are just a few of the salient changes in the structure of the rural economy that form part of this story.

In summary, this book positions peasants as central to the unfolding of the cotton revolution, in contrast to the dominant development narrative which paints them as passive recipients of new technologies and institutions, and which places foreign capital and the state at the center of discussion. It does not, however, pretend to present a peasant counter-hegemonic development narrative in which peasant voices dominate. Rather, this book seeks to grapple with the multi-layered negotiations and tensions among peasants and the multiplicity of "cotton interests," as well as between them, to capture the dynamics of agricultural intensification and agrarian change. This alternative story, in which African farmers' actions are constitutive of the institutional environment which in turn influences their activities, is a good example of how "the structural properties of social systems are both the medium and the outcome of the practices that constitute those systems."[8]

My interest in showing "how the past makes itself felt in the present" requires that we take a longer historical view.[9] This book begins, therefore, not at independence, as does the dominant development narrative, but in the late pre-colonial period in which colonial cotton policies took root. Subsequent chapters focus on the relatively short but important colonial period, when cotton policies failed to produce the quantities of cotton desired by merchant houses and metropolitan textile companies. It is only in the last two chapters that we examine the contemporary period, in which we witness both an intensification of cotton under favorable economic conditions (1970–84) and its extensification as conditions grew worse after the mid-1980s.

Cotton and the discourse of development

Development narratives by themselves tell us little about the social and historical contexts in which they are produced, nor why some stories become more dominant than others.[10] Recent scholarship on the discourse of development offers some guidance in answering these questions. In their analyses of the words and texts of development, scholars such as James Ferguson, Timothy Mitchell, and Arturo Escobar demonstrate how development agencies necessarily construct images and narratives of development that reduce complex social and economic dynamics to technical "problems" that fall within the purview of development planning. James Ferguson shows how the development agency texts on Lesotho, particularly World Bank project documents, construct the image of an autonomous national economy based on a subsistence agriculture that is "waiting to be developed."[11] The reality is that 70% of rural households' income is derived from labor migration to neighboring South Africa. "Acknowledging that Lesotho is a labor reserve for South African mining and industry rather than portraying it as an autonomous 'national economy', moreover, would be to stress the importance of something which is inaccessible to a 'development' planner in Lesotho."[12] Development agencies must therefore "rearrange reality" to present Lesotho as a nation of farmers, rather than one of migrant laborers, because agricultural development schemes lie within their jurisdiction. Similarly, the state is presented as a neutral entity whose sole aim is to provide social services and promote economic growth. Considerations of politics, class biases, and the socially differentiated rural households which mediate development projects are excluded from development texts. It is this discursive practice of constructing an object of development that falls within the purview of "developers" which inevitably silences subjects such as Lesotho's migrant workers, landless laborers in rural Egypt, and peasant cotton growers in Côte d'Ivoire.

In his study of the construction of Egypt in the development texts of international development agencies, Timothy Mitchell shows how peasants are conventionally depicted in ahistorical terms as tradition bound and incapable of transforming their livelihoods. In these texts, the catalyst for economic and social change invariably comes from outside experts promoting new technologies, institutions, and development strategies.[13] The image of stagnant agricultural systems is just one of many tropes of the development discourse: others include images of chaos, crisis, violence, insecurity, instability, and environmental degradation. As Jonathan Crush argues, "the language of 'crisis' and disintegration creates a logical need for external intervention and management . . . Development animates the static and manages the chaotic."[14] In contrast to this static view of peasants and farming systems, the state and foreign capital are portrayed as all-powerful actors possessing the capacity to define "the space of development . . . in which only certain things could be said or even imagined."[15] Arturo Escobar argues that this space or "perceptual domain" is the creation of a "vast institutional network" encompassing international organizations, research institutes, universities, and local development agencies. Like Ferguson, Escobar presents development as a hegemonic discourse. It "constructs the contemporary Third World, silently, without our noticing it."[16]

The great paradox of this conception of development is that it leaves little room for local actors and social and economic processes that might influence the form and nature of development interventions. Nor does it consider disagreements and tensions among bureaucrats, capitalists, and development experts in the realm of rural development policy. For Escobar, the development apparatus is an instrument of domination. Its power is hegemonic, with a capacity to determine what can be thought and said within the spaces which only it can define. The power of development is in its ability "to integrate, manage, and control countries and populations in increasingly detailed and encompassing ways."[17] As Jonathan Crush states, "power in the context of development is power *exercised*, power *over.*"[18]

By emphasizing the hegemony of western ideas and institutional structures and the unidirectional nature of power in development discourses, Escobar, Ferguson, and Mitchell alike give insufficient attention to the power of local processes to shape development policies and programs. Their analytical framework also elides the role of conflicts and disputes among policy makers in the design of rural development interventions. This book argues that the so-called subjects of development have been important agents whose actions (resistance, innovations, etc.) have repeatedly reconfigured the "space of development" and contributed in significant ways to the nature and direction of agricultural change. Peasant involvement in this story has been and continues to be inextricably linked to the multiple and often contradictory interventions of bureaucrats, capitalists, and agricultural development experts. Indeed, as we shall see, it has often been at the interstices of these interventions that farmers have found space to adjust to and influence agricultural policies. This case study contributes to the debate by showing that the cotton revolution has been the outcome of multi-layered negotiations among farmers, government officials, and a heterogeneous group of cotton promoters over time. Farmers alternately resisted and accepted its cultivation under conditions that were subject to repeated contestation and negotiation. By focusing on the arenas of contestation, innovation, and social and economic change, this book views agricultural change as a dynamic process and negotiated outcome.

Contrary to the ahistorical and instrumentalist view of "development" as determined by "outside" interests, this history shows how cotton policies were significantly influenced by past failures to transform peasant farming systems. Drawing on the work of Sara Berry on the dynamics of agrarian change in Africa, this study suggests that state intervention into rural economies has been "intrusive" rather than hegemonic.[19] Governments, international donors, and a diverse group of cotton developers have introduced changes in the conditions by which farmers gain access to productive resources, but they have not determined how these resources are ultimately utilized. Patterns of agricultural growth are affected by changing conditions of access to productive resources but not in a predictable manner. As Berry persuasively argues, the historical record reveals considerable variation in the pattern of agricultural intensification

> which has been neither inevitable nor continuous in African farming systems. In some areas, intensification was halted or reversed by changing environmental or political and economic conditions; in others, it has occurred not as an adaptive response to population growth or commercialization, but in the face of growing labor shortages and declining commercial activity. Such cases underscore the importance of studying farming as a dynamic social process.[20]

A comparative reading of the history of cotton in other regions of Africa suggests that the Ivorian case is not unique. Recent scholarship by social historians on cotton and colonialism in Africa demonstrates the importance of local social and economic processes in shaping the experience of cotton colonialism.[21] For example, case studies from Mali,[22] Togo,[23] Nigeria,[24] and the Congo (formerly Zaire)[25] show how peasant economic rationality and local handicraft industries frequently frustrated the often grandiose schemes of development planners to increase domestic production for export markets. This encounter between local and regional processes and the world economy led to complex patterns of conflict and agrarian change whose dynamics varied across time and space. Richard Roberts's study of cotton policies in colonial French Soudan (Mali) illuminates the tensions and contradictions that divided colonial planners, administrators, and agricultural experts in the area of cotton development. The disagreements over whether to support peasant grown cotton or large-scale irrigated projects managed by Europeans led policy makers to adopt shifting and contradictory policies which had significant repercussions on production. Roberts shows that the limited resources and capacity of the colonial state to intervene in African rural economies forced it to focus its interventions around the linked issues of cotton prices, quality, and labor in order to increase cotton exports. His social history reveals that these policies were influenced as much by ordinary African men and women as by colonial administrators, agricultural agents, and French textile interest groups.

The contribution of the Ivorian case study to this scholarship is to demonstrate both the multiple ways that farmers have limited the state's (and foreigners') capacity to control agricultural production and marketing, *and* how peasant practices have influenced the very agricultural policies and innovations that are cited as being instrumental to the cotton revolution. This does not mean that peasant ingenuity and resistance has been the motor of agrarian history, or that we can reduce this complicated agricultural history to a simple story pitting rational peasants against an oppressive state and foreign capitalists. First, the study points to changing patterns of social differentiation within households and communities which makes it difficult to speak of "peasants" as a single, harmonious group with identical interests. This story chronicles the struggles among and between generations, men and women, and prosperous and poor farmers, particularly in the post-World War II period. Second, although peasants have played a significant role in shaping agricultural policies and practices, they have done so under conditions which have not been entirely of their own making. The colonial and post-colonial states and their representatives, development experts, and cotton and textile companies have historically sought to direct farmers down a road of export-oriented agricultural intensification. The shifting and contradictory policies adopted by these different agents limited the state's capacity to control agricultural production and marketing. In sum, I argue that peasants have been key players, but by no means the only players, in the making of this West African agricultural success story.

This book also contributes to recent scholarship on the social history of cotton by examining the dynamics of cotton growing during both the colonial and post-colonial periods. Previous works have focused largely on cotton policies during the pre-World War II period, when output levels were often correlated with levels of coercion. When forced cultivation ebbed, cotton output declined. Cotton production in the post-World War II, and especially the post-colonial, eras reached unprecedented production levels under changing social, political, and economic circumstances. The Ivorian case study thus allows one to examine the conditions under which peasant farmers participated in the shaping of a new agricultural landscape.

In highlighting the historical involvement of peasant farmers in this agrarian transformation, I have two goals. The first is to present an alternative analytical approach to agricultural intensification in Africa that emphasizes what Jean-Pierre Chauveau describes as "the cumulative effects of incremental innovations" to agrarian transformation.[26] That is, changes in farming techniques and modes of social organization that have evolved over decades can play just as decisive a role in agricultural growth as the sudden introduction of a new technological package developed at an agricultural research center and diffused by extension agents. This social historical approach allows one to explain not only how and why a development trajectory took the course it did, but also why it shifted direction at different periods.[27] It requires that we consider the temporal and social dimensions to innovation, as much as the technological and institutional forms that an innovation assumes.

My second goal is to highlight the originality of this example of agricultural development in West Africa by showing the extent to which agricultural intensification in Côte d'Ivoire is a continually negotiated outcome. In contrast to the heroic view of agricultural change as "dramatic modernization"[28] that emphasizes the role of exogenous forces as progressive agents of change (crop scientists, extension agents, the state) and farmers as mere recipients of technological and institutional innovations, this study argues that the cotton success story is the product of the interactive effects of induced and directed innovations under changing social, cultural, agro-economic, and political conditions that have united and divided different communities since the turn of the twentieth century. In the following sections, I outline the distinctive elements of this agrarian transformation by showing how current theories of agricultural intensification fail to explain its dynamics in a satisfactory manner.

Defining and explaining agricultural revolutions

This book's focus on an agricultural revolution in Côte d'Ivoire requires that we define our basic terms. What is an agricultural revolution? What do I mean when I refer to the "cotton revolution"? What indicators can we use to measure rates of agricultural change? How reliable are these measures? If we can agree upon the answers to these basic questions, we can then move on to address a second set of questions focused on explaining the pattern of revolutionary growth. For example, what are the sources of growth? What are the proximate or farm-level factors as opposed to the off-farm institutional, economic and policy conditions which influence and are influenced by farmers? I will now turn to the first set of questions.

One way of defining an agricultural revolution is to measure rates of growth of agricultural output. An agricultural revolution could thus be defined as a period characterized by an extraordinary increase in the average annual growth rate of agricultural production. A second definition sometimes used by agricultural historians compares rates of increase in agricultural output against rates of population growth. An agricultural revolution defined in these terms would be a period during which growth rates in the volume of agricultural production substantially exceeded population growth rates.[29] There are problems, however, in using total agricultural production as the measure of growth. First, increases in total output can result from an extensification of agriculture as much as from intensification. Extensification generally refers to an expansion in cultivated area. It can also refer to situations in which farmers spread their labor and other inputs more thinly over a larger or smaller area. Intensification most often implies some technological change that involves greater use of labor or other inputs per land unit.[30] The dramatic increase in coffee and cocoa production in Côte d'Ivoire between 1950 and 1990 is an example of agricultural extensification. Yields were low (300–450 kg/ha) for both crops by international standards.[31] The expansion in total output of these two mainstays of the Ivorian economy has historically been the result of pioneer agriculturalists establishing new plantations in sparsely settled areas of the tropical rainforest.[32] The growth in output has been proportional to the expansion in newly cultivated areas. As land availability has declined and the pioneer front slowed, yields have declined and total production has fallen.

In contrast to coffee and cocoa, the growth in cotton production is an example of agricultural intensification. The upward trend in output levels is the result of regular increases in land productivity as measured by cotton yields per hectare. This example of agricultural intensification brings us to a third way of defining an agricultural revolution; that is, it is a period characterized by impressive increases in land productivity as measured by yields of individual crops. The cotton revolution in Côte d'Ivoire is thus defined as that period in which yields per hectare steadily increased at a rate that contributed significantly to agricultural output. Higher yields per hectare combined with an expanding area under cultivation were together responsible for rapid growth rates in total production. The indicators for the years 1965–84 furnish ample evidence for what I call a "cotton revolution." Cotton yields increased at an average rate of more than 4% per year over these two decades. Cotton area grew at an even faster rate of 17% per year (Fig. 1.1). The latter was due to a threefold increase in the number of growers and secondly to an expansion in cultivated area per planter. The latter nearly doubled, rising from 0.77 ha to 1.40 ha per grower.[33] Yield increases accounted for 15% of the increased output over this twenty-year period. Improved yields were linked to both social and institutional changes, as well as to intensive use of fertilizers, pesticides, and herbicides. Although it represents less than 10% of the national area in cash crops, cotton accounts for 60% of agricultural inputs in Ivorian agriculture.[34] In all, 85% of the growth was due to the expansion in cotton area. This expansion in cotton area was itself made possible by the rapid diffusion of ox-plows in the 1970s and 1980s. The average area cultivated by ox-plow owners in 1984 (2.69 ha) was more than double the average cotton area

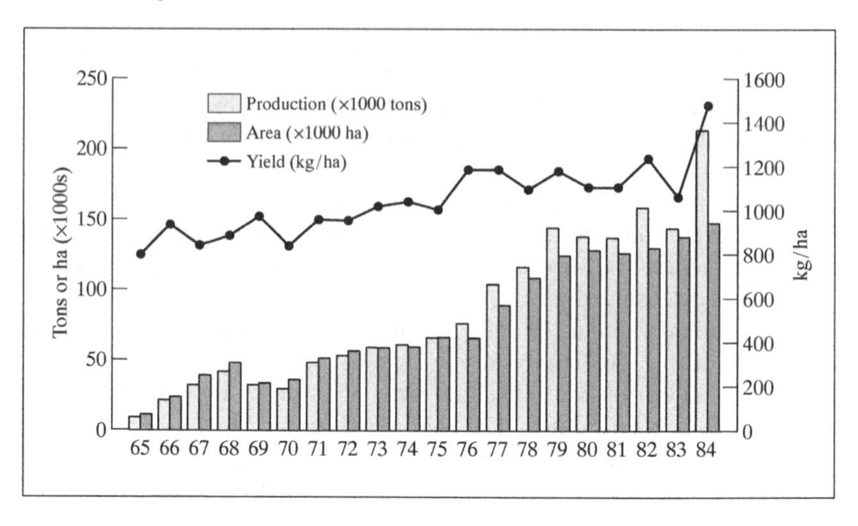

Fig. 1.1 Côte d'Ivoire cotton production, 1965–84 (*Source:* CIDT)

farmed manually (1.17 ha). Animal traction allows farmers to plow and sow their fields more quickly after the first rains.

How reliable are these measures? I argue in chapter 5 that the numbers are generally dependable due to the control of the vertically integrated cotton sector by CFDT and La Compagnie ivorien pour le développement des fibres textiles (CIDT) (Ivorian company for the development of textile fibers). The company's extension agents reside in villages and systematically record the number of growers, area cultivated, input distribution, and total production. CIDT's historical control of cotton markets, input supply, and credit enabled the cotton company to monitor these basic production indicators. However, the quality of the data declined after 1984 when CIDT modified its extension system and input subsidies were withdrawn at the behest of the World Bank. After this date, an extensification of cotton began in earnest as farmers spread more costly inputs over a wider than recommended area. Farmers typically under-reported the area in cotton to CIDT extension agents. A growing number of women also began to cultivate cotton but did so without the knowledge of CIDT. As a result, there is a discrepancy between cotton company figures and my own data on the number of growers and area cultivated in cotton. I find that CFDT/CIDT's data seriously underestimate both the number of planters and the area under cultivation which means that its yields are inflated.[35] Given the company's stake in demonstrating that its development model and interventions were successful, the exaggerated yields after 1984 are not surprising.

How can we explain the cotton revolution? Put differently, what are the sources of productivity growth?[36] There is an abundant literature on agricultural intensification that offers a variety of possible explanations. Brush and Turner review the major theories of agricultural change which they group into organizing themes.[37] These include technology themes (neo-Malthusian and technological diffusion theories), demand themes (Boserupian and Chananovian theories), and political economy themes (dependency, modes of production theories). One of their most challenging conclusions is that any explanation of agricultural change must consider "an array of material variables (tools, cultivars, labor, capital, soil, climate), structural variables (national and international economy, local institutions, social organizations, and responsibilities), and individual behavior (goals and allocation choices)."[38] This book seeks to provide such a complete explanation of the cotton revolution by showing how all three variables (material, structural, and behavioral) have interacted in different combinations over time.

The interplay of induced and directed innovations

Vernon Ruttan summarizes the most influential agricultural development models of the post-war period up through the 1970s. He focuses in particular on the failure of most models (e.g. diffusion, and high-payoff input models) "to explain how economic conditions induce the development and adaptation of an efficient set of technologies for a particular society."[39] Ruttan and Hayami's influential induced innovation model posits that technological innovation and agricultural growth result from farmers pressuring public and private research institutions to develop new technologies to help them overcome relative scarcities in the factors of production.[40] This bottom-up theory of agricultural innovation has been soundly critiqued for failing to specify the nature of the social, economic and political circumstances in which the demand for innovations take place.[41] In response to their critics, Ruttan and Hayami acknowledge that their model lacks a "theory of action";[42] that is, it fails to show how farmer-research institution relationships are structured. Bruce Koppel suggests there is a "complex political economy dimension" behind these relationships.[43]

John Staatz and Carl Eicher's review essay on agricultural development theories covers much of the same terrain as Ruttan's overview but takes the discussion up through the 1980s.[44] Of interest to this book is their notice that development economists continue to focus on the economic and technocratic aspects of development while ignoring the issues of power and politics that most often drive the process of agricultural change. Larry Burmeister's model of "directed development" explicitly addresses the links between social power and agricultural research and extension. It is also the one that helps to describe the dynamics of technological change and agricultural growth in Côte d'Ivoire.[45] In contrast to the induced-innovation model of agricultural development, he shows that the direction of technological and institutional change in South Korea's Green Revolution was driven by the state's political objective of achieving national self-sufficiency in rice. Yield maximization goals of the "agrobureaucracy" overrode farmer concerns about integrating the new high-yielding variety into local farming systems and the HYV's tolerance to cold weather and pests. In short, peasant farmer input into the process of agricultural research and extension was negligible in the South Korean Green Revolution.

While more successful in specifying the political and economic conditions in which government research and development programs operate, the political economy critique of induced innovation does not leave much room for peasant agency or for conflicts among state agencies and development professionals, and policy makers. For example, the power of peasant farmers to influence agricultural research and extension is severely limited in the directed development model. Turning induced innovation theory on its head, Burmeister's farmers are induced by the state to utilize resources in ways that are congruent with its policies.

> There was little proactive input by farmers into the production and dissemination of the new technology. The economic and organizational fields in which farmers were constrained to make technology adoption decisions were created and/or manipulated by state agencies. Microlevel decisions were in effect channeled to coincide with state managers' macrolevel national development plans.[46]

Burmeister admits that the high degree of state autonomy in South Korea probably represents an extreme case along a continuum of state/society relationships.[47] He thus leaves open the possibility that agricultural change may be driven by both locally induced (endogenous) and externally directed (exogenous) forces, a dynamic which comes closest to describing the development of cotton in Côte d'Ivoire. For example, one can identify a set of top down, externally directed technological and institutional innovations such as the high-yielding cotton package and the vertically integrated CFDT system and a set of bottom-up or induced innovations such as changes in farming practices and the social organization of production which facilitated the integration of the high-yielding package into the farming system. However, the social and historical origins of these ostensibly exogenous and endogenous innovations suggest that they do not neatly fit into these ideal categories. As Jean-Pierre Olivier de Sardan argues, innovations are commonly hybrid constructions which *emerge* from the interactions of different social groups at specific times and places.[48] The innovations pertinent to the Ivorian cotton revolution gestated in the numerous negotiations and conflicts bringing various agents together around the issue of export-oriented cotton growing. They were born and subsequently modified, sometimes disappearing altogether, in the context of these encounters.[49] In short, in order to capture the social and historical dimensions of technological and institutional innovation and to understand why cotton took off in the middle 1960s and not earlier, we need to transcend the heuristically useful but analytically limited concepts of exogenous and endogenous forces to consider the multiple encounters and incremental changes in society and economy in which the seeds of the cotton revolution took root.[50] To emphasize the interdependency of structure and agency in which agricultural innovations are not purely external nor internal to different social groups but are the outcome of dynamic social and agro-ecological processes, I will refer to the *interplay* or *interactive effects* of exogenous and endogenous forces in this study.[51]

Agricultural development and agrarian politics

This conceptualization of farmer agency in the shaping of a West African cotton revolution contrasts with the populist argument of Paul Richards. In his book *Indigenous Agricultural Revolution*, Richards argues that most agricultural development projects in West Africa have failed because they have not involved farmers in the problem identification and project design phases. Like Burmeister, Richards views the technological innovation process as being driven by external actors, notably rural development planners and agronomists working at experimental research stations, who seek to replace rather than build upon indigenous farming practices. Richards shows that peasant farmers are continually experimenting with new crops and farming techniques to solve "location-specific, ecologically particularistic" problems but that extension agents are oblivious to these local experimental initiatives.[52] He describes the coexistence of such formal and informal research and development (R and D) efforts as "two systems (that) pass like ships in the night."[53] According to Richards one of the great challenges for contemporary agricultural development planning is to bridge this gulf between peasant farmers and extension agents through decentralized, "bottom up" participatory approaches to rural development in which farmers are involved in problem identification and project design and agricultural experts become collaborators and consultants.[54]

In Richards conceptualization of agricultural development, directed-development is a dead end. Even when externally driven development schemes are oriented towards small farmers, he points to minimal farmer participation, high drop out rates, and inappropriate technology as a basis for skepticism.[55] Farmers are portrayed as resistant to major modifications of their farming system, such as replacing intercropping with monocropping, due to the associated risks of increased labor demands, unreliable crop yields, and increased vulnerability to pest, weed, and disease problems.[56] Their conservatism is paradoxical in light of the emphasis given by Richards to farmer experimentation and innovation. This rigidity is reinforced by the oppositional terms in which peasants and agricultural development agencies are drawn: bottom-up/top down; indigenous/external; informal/formal; particularistic/general. There is little room here for dialectical relationships, hybrid forms, intermediate categories, and socially differentiated actors that bridge and transform these seeming opposites.[57] Nor is there much consideration of the possibility that "exogenous" innovations might offer a measure of security to peasant farmers (e.g. drought tolerant cotton varieties; guaranteed producer prices, access to credit) and thus be viewed as desirable.

One reason for this ambiguous portrait of peasants is that Richards's agrarian populism eschews a consideration of politics, social organization, and rural economy.[58] As a result, his conceptual framework cannot accommodate one of the most significant agricultural success stories in late twentieth-century West Africa. For example, despite the agro-ecological wisdom of intercropping cotton with food crops, Senufo farmers abandoned this practice under changing social and agronomic conditions during the 1960s. In fact, intercropping in general, has declined dramatically as farmers have adopted new farming techniques (ox-plows, herbicides) and high-yielding varieties. According to Richards's analysis, this abandonment of time-tested and location specific agro-ecological practices for higher risk and uniform technological packages was an unlikely outcome.[59] This book explains how and why such changes in indigenous farming practices came about by locating them on the larger canvas of agrarian change in which cotton growing has transformed the agricultural landscape.

This study's focus on the capacity of large numbers of small farmers to influence agricultural policies is also of interest to scholars of African agrarian politics. Robert Bates's work on the political basis of agricultural policies emphasizes national-level political and economic considerations in the setting of producer prices in input and output markets. His rational choice political economy perspective views small farmers as particularly disadvantaged when it comes to influencing agricultural policies through collective action.[60] For example, Bates argues that "when producers are numerous and widely scattered, the costs of organizing are higher."[61] Large farmers are considered to have greater influence, in part because the costs of organizing and lobbying are lower and the problem of "free-riding" is minimized.[62] Peasants are also considered to be fearful of government reprisals when they organize to defend their collective interests. Rather than form political parties or openly protest adverse pricing policies, Bates argues that peasants more typically "use the market against the state."[63] By this he refers to the tendency of peasant farmers to avoid worsening agricultural market conditions by reducing the area under cultivation of economically unattractive crops, investing their limited resources in more remunerative crops, selling their produce in parallel markets, and by emigrating to other areas to obtain higher incomes. The case of commercial cotton growing in Côte d'Ivoire confirms this rational choice perspective. Peasants have historically used the market in these and other innovative ways to defend themselves against adverse price changes in input and output markets with the intended and sometimes unintended consequences of producing more favorable conditions.

Cotton growers have also used what James Scott calls "everyday forms of resistance" to buffer themselves against adverse market conditions and the depredations of the state (e.g. excessive taxes, forced labor).[64] That is, rather than engage in collective action and risk reprisal, peasants commonly engage in more covert and individual activities such as foot-dragging, false compliance, and sabotage. Allen Isaacman's analysis of cotton growing in colonial Mozambique reveals a large repertoire of peasant resistance tactics to the Portuguese cotton regime.[65] In Côte d'Ivoire, peasant opposition to forced cultivation often took the form of crop neglect (late sowing and weeding), sabotage (dumping fertilizer), and lack of compliance with cropping standards (field locations, harvest times). In recent years under the CFDT system, farmers have minimized the negative effects of low market prices by defaulting on debts, bribing cotton graders to classify their low-grade cotton as high grade, and by selling subsidized crop inputs (seeds, fertilizers, pesticides) rather than using them in their own fields. Like using the market against the state, such actions are typically undertaken by individual farmers and require little organization.

The Ivorian materials suggest that both Bates and Scott underestimate the capacity of large numbers of small farmers to organize collectively to claim what they perceive as their rightful share to resources. That is, peasant farmers do not simply "dodge and maneuver" in response to the requisitions of the state and foreign agribusiness in the rural economy.[66] In Côte d'Ivoire, they have also engaged in direct political action.[67] The example of the 1991 cotton market strike organized by peasant producer organizations discussed in chapter 6 illustrates the capacity of small farmers to defend their interests through collective action. The ability of farmers to mobilize around this strike was enhanced by the space opened up by multiparty politics as well as through the evolving institutional structure of farmer cooperatives (Groupements à vocation coopérative [GVC]). These small farmer organizations minimized the effort expended in coordinating farmers to withhold cotton from CIDT which resulted in more attractive input prices that benefited all cotton growers. In Bates's analysis, this was an unlikely outcome. The Ivorian case study shows how and why small-scale cotton growers have emerged as an increasingly powerful lobby that has influenced agricultural policy in the 1990s.

In summary, the agrarian politics and populist arguments of Bates, Scott, and Richards enhance our understanding of the myriad ways that peasants farmers manipulate social, ecological, and/or institutional processes to their advantage. Given the sometimes overwhelming odds weighing against them in the market place and the claims made on their limited resources by a panoply of agents, peasants are portrayed as largely conservative actors who tend to keep their distance from what are perceived to be high risk situations. As Richards puts it, "many small-scale farmers prefer to minimize risks rather than maximize output."[68] While I agree with Richards that risk-aversion is a concern of peasant farmers working poor soils in areas of unreliable rainfall, it has not stopped them from abandoning "traditional" farming practices like intercropping and taking up high-input and labor intensive cash crops in monocropped fields as tens of thousands of cotton growers have in Côte d'Ivoire. Peasant farmers have also displayed a remarkable ability to engage in political action in ways that neither Bates nor Scott would have anticipated.[69] The interest of the Ivorian cotton story is that its central characters, peasant farmers, assume roles that scholars of African agrarian studies have not sufficiently considered. The following sections provide more background on the research locations and methods that I have used to gather the necessary materials to tell this story.

Research site

The geographical focus of this study is the Korhogo region of northern Côte d'Ivoire (Fig. 1.2). This area is at the center of the country's major cotton growing area (Korhogo-Ferkéssédugu-Boundiali-Mankono) where 60–70% of the national seed cotton is produced.[70] In 1995–96, the Korhogo region alone accounted for a quarter of Côte d'Ivoire's total seed cotton production. During the colonial period, colonial officials, the French textile industry, and merchants hoped to transform the area into a major cotton producing zone. The colonial archives document the successive and invariably unsuccessful steps taken to realize this objective. Today, northern Côte

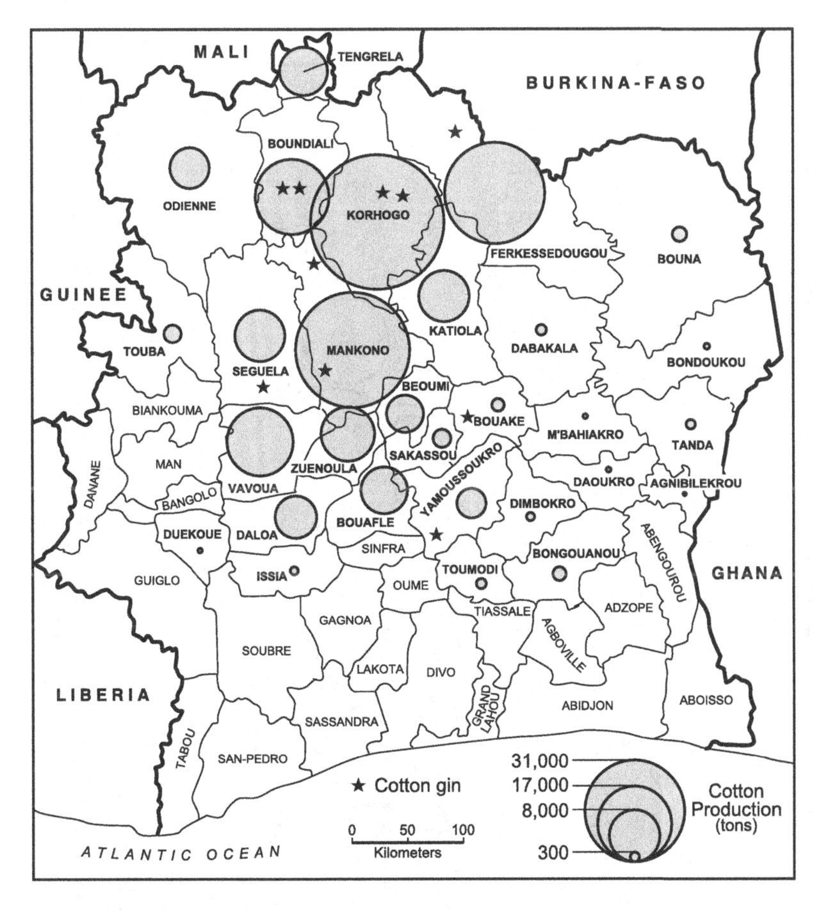

Fig. 1.2 Reference map of Côte d'Ivoire's cotton growing areas, 1995–96 (*Source:* CIDT)

d'Ivoire, southern Mali, and southwestern Burkina Faso, are major producers in the West African cotton belt (Fig. 1.3).[71]

In order to understand the social and agricultural dynamics of cotton growing within this zone, I systematically collected data in the community of Katiali, located in the subprefecture of Niofoin in the Department of Korhogo. A village of some 1,800 persons, Katiali is inhabited by Senufo- and Jula-speaking peoples. It was the center of a pre-colonial chiefdom that was progressively dismembered as a result of warfare with the neighboring M'Bengué chiefdom and encounters with the armies of Babemba Traoré

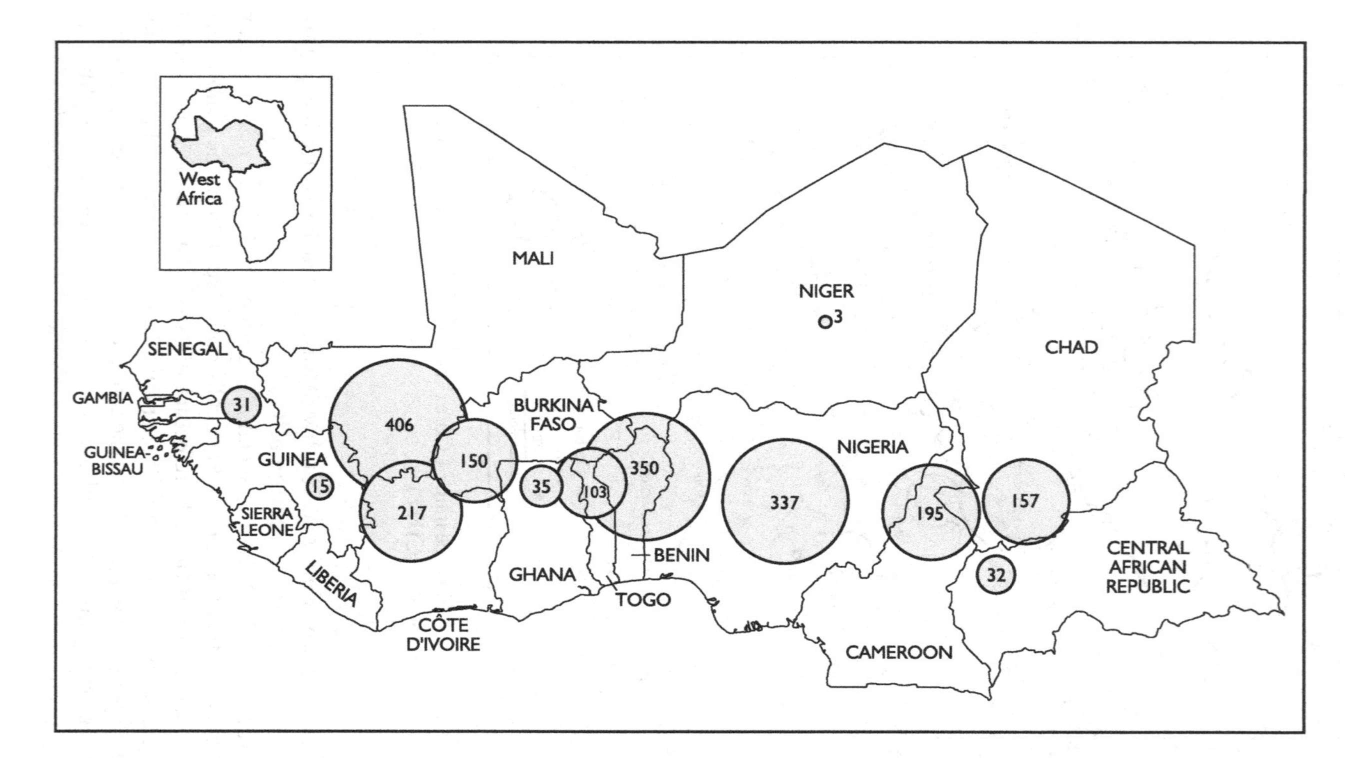

Fig. 1.3 West African Cotton Production – 1996 (in thousands of tons of seed cotton) (*Source:* CIDT, FAO)

and Samori Touré in the late nineteenth century. It is located 60 kilometers northwest of Korhogo, in the transition between the Guinea and Sudanian savannas.

Situated in the sub-humid tropical zone between eight and eleven degrees north latitude, the climate of the Korhogo region is characterized by distinct rainy and dry seasons. Rain falls within a five to six month period (May–October), averaging between 1,000 to 1,200 mm annually. The dry season begins in October and ends in early May. The rhythm of wet and dry seasons is regulated by the NE-SW movement of air masses within the Inter-Tropical Discontinuity Zone.[72] Hot, dry, "Harmattan" winds blow south from the Sahara during the months of January and February.[73] The intensity and concentration of rainfall in the months of July–August (22 mm/rain a day) gives the region's climate an aggressive character. Deep chemical weathering and the process of laterization produce ferruginous duricrusts and ferrallitic soils (red latosols) with low soil fertility.[74] Depending on the density of the plant cover and agronomic practices, soil erosion can be a serious problem.[75]

The major ethnic group in the study area is the Senufo. Although the Senufo inhabit parts of southern Mali and southwestern Burkina Faso as well as the north central region of Côte d'Ivoire, this study focuses on the Senufo of the Korhogo region (Fig. 1.4). Within the Korhogo region itself, the Senufo are composed of a number of subgroups based on differences in dialects. The Senufo are primarily farmers and animistic in faith. Mande-speaking Jula are the second most important ethnic group in the area. Noted in the past for dominating local and long-distance trade in the Korhogo region and beyond, the Jula today are considered to be "traders without trade."[76] For the most part dispersed and living in enclaves among the Senufo, they primarily engage in farming and weaving.

Since the late 1960s, a growing number of Fulbé herders have entered the region from the neighboring countries of Mali and Burkina Faso.[77] According to the former Ivorian livestock development agency (SODEPRA), there were approximately 37,000 Fulbé cattle in the country in 1989. Kientz estimates that the number of Fulbé cattle grew at an impressive annual rate of 9% between 1968–89.[78] There is a marked concentration of Fulbé herds in the north central region.[79] Relations between the largely transhumant Fulbé and the local population continue to be extremely tense due to the problem of uncompensated crop damage caused by Fulbé herds in farmers' fields.[80]

The major staples grown in the mid-1990s were maize, rice, peanuts, and sorghum. These four crops, planted singly or in combination, accounted for 66% of the total cultivated area in Katiali. Cotton covered 33% of the area. Yams are an important staple in the southern Korhogo

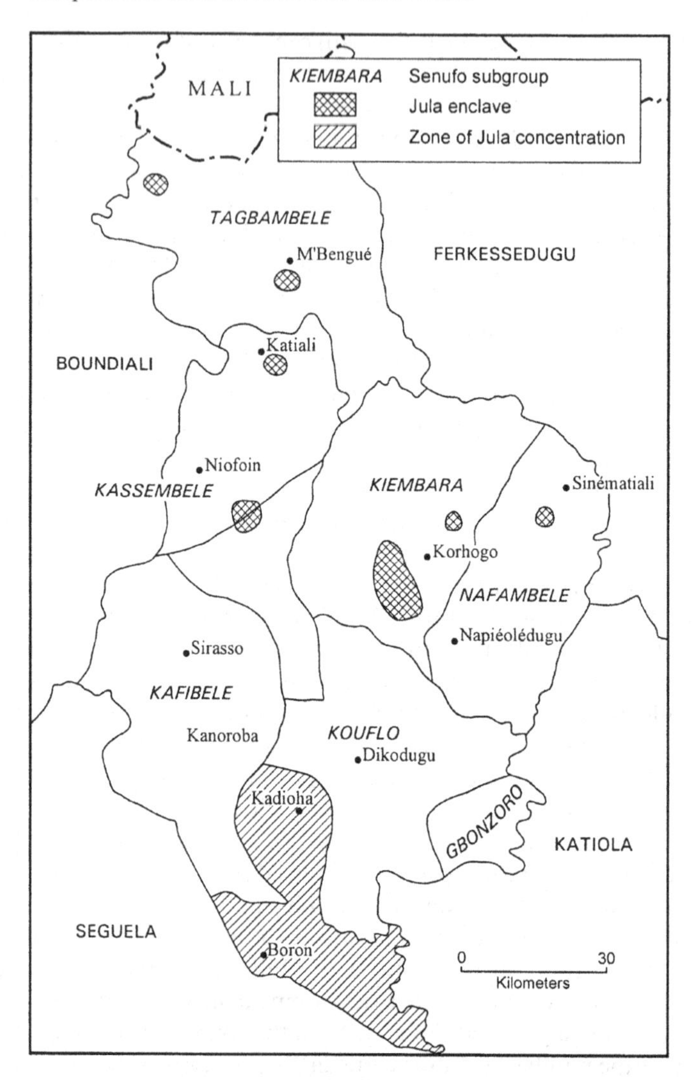

Fig. 1.4 Distribution of Senufo subgroups in Côte d'Ivoire (derived from Coulibaly, *Le paysan*, p. 53).

region. At present, no land is privately held outside of urban areas in the region. Customary land use rights based on usufruct tenure still prevail. With the exception of a 15–20 km radius around the city of Korhogo, where population densities average eighty persons per square kilometer, population densities rarely exceed twenty inhabitants per square kilometer in the region.[81]

Research methods

Reconstructing the history of agrarian change is fraught with epistemological difficulties, as most field researchers will acknowledge.[82] The challenges are multiple and vary according to the research locale. Such challenges include the relative availability and reliability of agricultural production statistics, the biases and gaps in colonial archives, the researcher's knowledge of local languages, and the extraordinarily subjective manner in which knowledge is produced. The assumptions and theoretical orientations I carried with me to Côte d'Ivoire for the first time in January of 1981 certainly influenced how I went about doing research. For example, my initial interest in studying the history of cotton stemmed from a concern about hunger and its relationship to competition between food crops and cash crops, colonialism, and rural capitalism.[83] Consequently, I conducted a year-long farm management study to better understand the place of cotton within the local farming system and economy.[84] My interest in the agricultural and social history of cotton drew me to archives in four different countries. In the National Archives of Côte d'Ivoire, I spent months reading the quarterly reports of local administrators. These documents, authored by the administrative chief officer (known as the "Commandant de Cercle"), contain the richest details on social and agricultural change from the perspective of colonial officials. Between 1898 and 1904, northern Côte d'Ivoire was administered by French military officers whose headquarters were in Bamako, in present-day Mali. Military commanders sent reports of their operations and administrative work to their superiors in Bamako. In June of 1982, I consulted these documents at the National Archives of Mali in Koulikoro.

In the French colonial administrative hierarchy, Dakar, Senegal, was the seat of the governor general of French West Africa (AOF). Every year the governors of each colony sent their annual reports to the governor general. Although not as detailed as the circle commander's reports, these annual reports contain information on policies and issues that had important repercussions at the local level. I consulted these and other useful documents at the National Archives of Senegal. The governor general, in turn, submitted his annual reports to the Minister of Colonies in Paris. Policy issues affecting all of French West Africa can be found in these documents, which were originally housed at rue Oudinot in Paris and are now found in the French National Archives, Overseas Section, in Aix-en-Provence. As will be evident in the early chapters of this book, much use was made of these archival materials in reconstructing the pre-World War II history of cotton.

For political reasons, the Ivorian national archives are closed for the post-World War II period. However, the Côte d'Ivoire Chamber of

Commerce located in Abidjan possesses an important archive on certain sectors of the post-war colonial economy. Thanks to the cotton sector dossiers, I was able to reconstruct colonial cotton policies up to independence in 1960.

Oral histories form a second cornerstone of this book. I conducted interviews with village elders, cotton extension agents, and cotton company officials throughout the 1980s and much of the 1990s on the topics covered in this book. After I had gained their confidence, men and women spoke openly on such issues as pre-colonial settlement and trade, the social organization of production, forced cultivation, migration, land rights, and farming systems. Throughout I have tried to balance archival accounts with information obtained from these oral sources, in order to provide as accurate a picture as possible of the transformation of peasant society and economy in the study area. The oral accounts of now retired cotton extension agents more than made up for the gap in archival sources for the 1950s. Their stories of forcibly introducing new cotton varieties in the 1950s and 1960s provided precious details into this important but little known chapter in the cotton revolution.

While conducting a twelve-month long farm management study with a small sample of seven households, I decided to administer a questionnaire to a larger sample of thirty-eight households stratified by local measures of relative wealth (e.g. number of oxen, active workers), to gain a more representative view of local farming systems. I collected information on the crops and area cultivated, agricultural techniques, off-farm labor, field locations, and land-use rights from a sample of nineteen Senufo and nineteen Jula households. I repeated this farming systems survey during subsequent visits to Katiali in 1986, 1988, 1992 and 1995. This longitudinal study provides the basis of my observations on the dynamics of farming practices over the 1980s and 1990s, which are summarized in chapters 5 and 6. As a result of these repeat visits and surveys, I began to appreciate not only the speed at which agricultural change had taken place, but also the direction and sometimes short-lived nature of these changes. The rise and decline of women cotton growers is a good example of such rapid changes. These changes would not have been discernible had I decided to follow the more conventional re-study approach, which commonly entails returning to the same research site ten to twenty years after the original study. The method of repeat visits and surveys allowed me to record these trends and the circumstances in which they evolved. My findings often ran counter to the views of agricultural stagnation that prevailed in the development discourse pertaining to African agriculture.[85] They also stood in contrast to the passive image of peasants entrapped in exploitative production relations with foreign capital and African states.[86] The results of this longitudinal study reveal a far more complex, dynamic, and interesting story. The information collected from these surveys constitutes the third cornerstone of this study.

Finally, I have accumulated agricultural statistics on cotton production trends from local, regional, national, and international sources. Locally, I have cross-checked my household surveys on cotton area and production with the records of village extension agents and cooperative leaders. At the district, regional, and national levels, I have relied on the Ivorian cotton company's reports and interviews with company officials residing at these different locations. I also made three visits to the headquarters of the French cotton firm CFDT in Paris, where I was welcomed to consult documents and discuss cotton development issues with company officials.

The photographs illustrating this book were obtained from two principal sources. I purchased the postcards illustrating indigenous and forced cotton marketing in Côte d'Ivoire from specialist dealers in Paris, France. The majority of the photographs I obtained from the photothèque of the Institut Fondamental de l'Afrique Noire (IFAN) in Dakar, Senegal. Although typically focusing on a limited number of themes, these images provide precious details that neither the oral histories nor written archival documents offer.[87]

Regarding the day-to-day routines of village-level field research, I regularly worked with an interpreter when conducting interviews in Senufo and Jula. My proficiency in Jula was adequate for conducting the farming system survey but fell short for the more qualitative and less formal interviews focused on cultural, historical, and agro-ecological topics. I have been fortunate to have been able to work with the same interpreter, Mr. Adama Koné, for more than fifteen years.

The general argument and organization of the book

From the very beginning, colonial cotton policies were influenced by local patterns of production and trade. Cotton was an important cash crop in the pre-colonial economy and fed into a thriving regional cotton handicraft textile industry. Chapter 2 examines the fragmentary evidence indicating the place of cotton in the rural economies of the Senufo and Jula communities of the upper Bandama River valley. The traditions of cotton growing, spinning, and weaving were of great interest to colonial administrators, who during the first decade of colonial rule sought to expand cotton production for the benefit of France's textile industry. The persistence of this handicraft industry during the colonial period was to have important repercussions on trade patterns, cotton prices, and colonial polices.

Cotton varieties play a central role in this agricultural, social, and economic history. The extent to which local cottons were acceptable in terms of yield and fiber quality (color, gin yield, length, and strength) was of central concern to colonial administrators, agronomists, merchants, and the French textile industry. This chapter examines the types of cotton cultivated, their origins, and how perennial cotton was grown. It demonstrates that the so-called "exotic" cottons favored by Europeans were in fact derived from a subspecific variety of wild African cotton. Moreover, there is good evidence showing that African cotton growers continually experimented with new varieties and hybrids, some of which influenced the direction of colonial cotton development programs.

Chapter 2 also documents the extreme social, economic, and political instability that characterized the last quarter of the nineteenth century. The conflicts generated by the expansionist policies of both African and European empire builders resulted in considerable hardships, social and political upheavals, and new settlement patterns on the eve of colonial rule. The demographic impact of these tumultuous times is hinted at in the observation of the French explorer/military officer Louis Gustave Binger, that population densities in the Niellé region declined from forty to fifteen persons per square kilometer during this time.

The failure of the colonial state to capture a large part of local cotton production is the focus of chapter 3. In addition to a host of labor supply, agronomic, and technological problems, the colonial state and commercial firms faced stiff competition in local markets. Jula merchants supplying the indigenous handicraft industry invariably offered higher prices than European merchant houses to cotton growers. Moreover, when food crop markets were buoyant, peasants devoted their limited resources to growing food crops like rice, yams, and maize. Peaks in cotton output mirrored heightened administrative coercion and hardships in the north. Disagreements among colonial officials over how best to stimulate cotton production, combined with the cotton quality criteria of the French textile industry and peasant responses to the regime of requisitions, led to major shifts in colonial economic and agricultural policies. The agrarian history of cotton over the period 1912–46 shows that the parallel cotton market played an important role in the design of the new cotton program that emerged over the following decades.

Colonial administrators and textile industry representatives recognized that the best way to deal with the parallel cotton market was to increase total output. They agreed that higher producer prices and production levels would lead to a win-win situation in which both local and export markets could be satisfied. Cotton development policies in the post-war period thus centered around the development of a high-yielding cotton package made possible by the development of new hybrid seed varieties and pesticides such as DDT. The new cotton program also included institutional reforms in which the French cotton company, CFDT, assumed a vertically integrated form. In addition to supplying growers with the new cotton package, CFDT gained the exclusive right to purchase the new *Allen* variety. This market monopsony effectively eliminated Jula merchants from cotton markets. The "CFDT system" was thus partly shaped by the evolving policy discussions during the late colonial period on how to compete with the parallel cotton market and at the same time produce a quality fiber that would sell in export markets. This chapter shows that peasant reluctance to participate in the new cotton program led to the return of coercion in the north during the first decade of "independence." The fact that peasants were forced to cultivate the new variety testifies to the directed nature of the process of technological innovation.

Chapters 5 and 6 are based on village-level longitudinal research, which allows for a more fine-grained analysis of the dynamics of agricultural intensification and extensification. Chapter 5 argues that the intensification of cotton resulted from the combined processes of directed and induced innovations. As noted above, cotton company research and development was biased in one way (land-saving), while innovations at the local level were biased in another (labor-saving). This chapter focuses on the range of locally induced innovations aimed at reducing labor bottlenecks and increasing individual household incomes. I argue that the emergence of new technologies, cropping patterns, and new social and institutional arrangements at the village level have been the key dynamics of the cotton revolution. These innovations effectively altered the structure of the rural economy and society in northern Côte d'Ivoire. The extension of credit tied to cotton production and marketing, along with the expansion of ox-plows and cultivated area, created new economic and labor supply problems whose solution involved new social relations between seniors and juniors and between men and women, as well as between relatively prosperous and poor households.

Chapter 6 shows that the pattern of agricultural intensification characteristic of the 1970s and early 1980s shifted to a pattern of extensification and diversification over the period 1985–95. The worsening economic circumstances brought on by the elimination of the fertilizer and pesticide subsidies, falling producer prices, a more demanding cotton variety, and currency devaluation forced farmers to reallocate their limited resources in new ways. The case of Katiali shows that the most common coping methods included the extensification of cotton, defaulting on loans, crop diversification, organized strikes, and the creation of new input supply and marketing cooperatives. The outcome of these new patterns of resource use was a precipitous fall in cotton yields and total production. To reverse this trend, the cotton company abandoned its new glandless variety, increased producer prices, and offered new subsidies to cotton growers. For the first time, cotton company officials in Bouaké and Paris acknowledged the power of peasant producers in shaping cotton policies. Nevertheless, cotton growers continue to be portrayed as passive subjects of development in the cotton development discourse of the late-1990s. In the conclusion, I summarize the main points of my study and highlight its contribution to the specialist literature on agricultural development in African agrarian studies.

2 The collision of empires, 1880–1911

> There are too many ruins to note them all on the map.
>
> Louis Gustave Binger, 1892

The late nineteenth century was an extremely violent period in the history of northern Côte d'Ivoire. Between 1880 and 1900 the area was successively invaded by both African and European armies that sought to incorporate it into their own spheres of influence. In the 1890s alone, the Korhogo region was occupied by troops of three different empires. Between 1883 and 1894, the area formed the southernmost portion of the Kénédugu empire, based in Sikasso. In 1894 Samori seized control of the region and incorporated it into his new eastern empire (Fig. 2.1). After their defeat of Samori in 1898, the French occupied the area and governed it under colonial rule until independence in 1960. The dramatic political changes of the late pre-colonial period were to have far-reaching economic and social repercussions during the colonial period.

This chapter highlights the transformation of the rural economy of the Korhogo region on the eve of colonial rule. It begins with fragmentary views of the pre-colonial cotton economy and indigenous handicraft industry provided by European explorers who passed through parts of the region in the early and late nineteenth century. This is followed by an overview of cotton growing in the Sudanian savanna of West Africa in the late nineteenth century with emphasis on the methods of cultivation and the varieties commonly planted. The chapter ends with an overview of the establishment of French colonial rule in northern Côte d'Ivoire, emphasizing how settlement patterns linked to the turmoil of the late nineteenth century influenced the nature of the administrative apparatus and the manner in which it intruded into rural areas.

European accounts of pre-colonial northern Côte d'Ivoire

Any attempt to reconstruct the rural economy of pre-colonial northern Côte d'Ivoire, particularly its agricultural systems, is made difficult by the

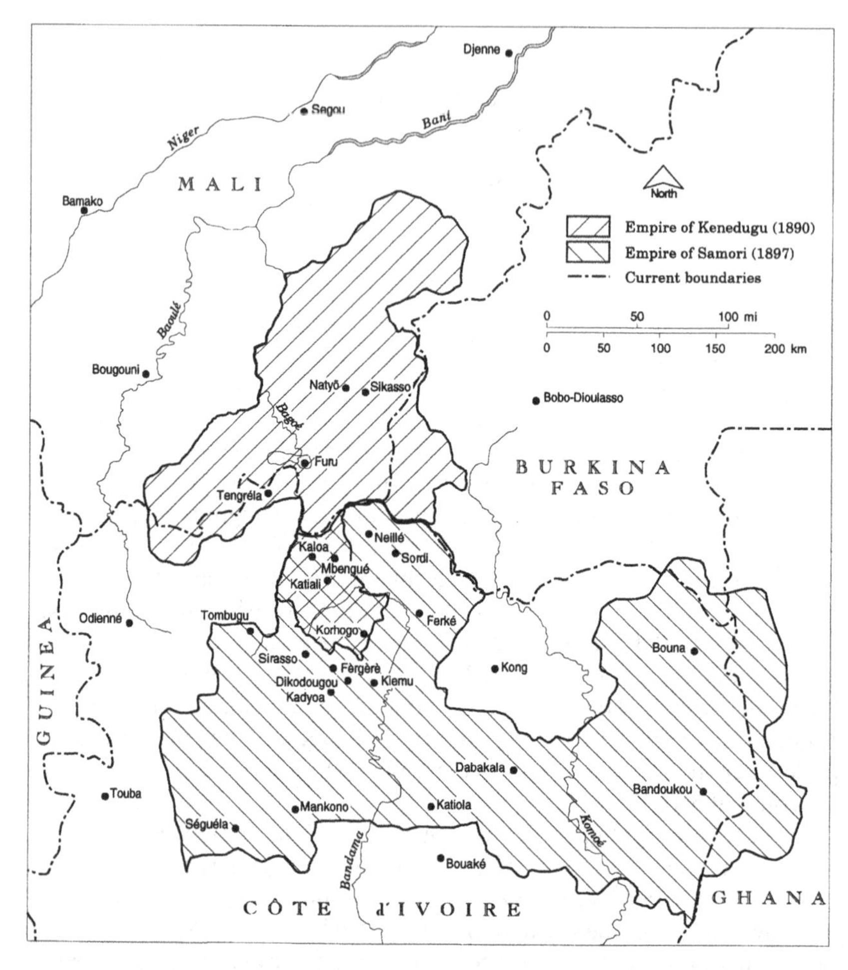

Fig. 2.1 Map of Kénédugu and Samori empires (derived from Person, *Cartes Historiques*, 1990, plates 33 and 41).

paucity of written documents and oral histories. When I began this research in 1981, only a few elders in Katiali could furnish information on the late pre-colonial period. Written documents dating from the pre-colonial period are equally scarce. Few European explorers visited the area prior to its colonization in 1898. For most of the nineteenth century, Europeans believed that a high mountain chain known as the Kong Mountains stretched across much of West Africa. This impressive but fictitious mountain range was believed to be a major obstacle to commerce and travel. It was not until Binger's exploration of the region in 1887–88 that the Kong Mountains were removed from maps and the area opened up to European conquest.[1]

The first European to enter the region was René Caillié who set out from Sierra Leone in 1827 disguised as a Moor to visit the fabled city of Timbuktu.[2] He passed through the Malinké areas of Sambatiguila and Timé before entering Senufo country south of Tengréla. Traveling in trade caravans led by Jula traders, Caillié tended to see the land and life of the areas he passed through from the perspective of his Jula informants. For example, he failed to identify the Senufo as a distinctive ethnic group; he simply called them "Bambara," the Jula word for pagan which was commonly used to identify non-Muslim groups.[3] Caillié's observations are, nevertheless, interesting and useful for their insights into political and economic life in the early nineteenth century. He noted that the Senufo were organized into a multitude of small chiefdoms, some of which were at war with each other. However, in contrast to Binger's account sixty years later, which depicted a countryside left in ruins as Samori and Tiéba expanded their empires, Caillié paints a peaceful picture of rural life. The area was rich in food crops: yams, maize, millet, peanuts, and rice "grow in abundance in this happy country."[4] Senufo farmers grew cotton which they exchanged with Jula men and women for salt and kola nuts.[5] Jula women spun raw cotton into thread which their husbands wove into cloth. On long-distance trade expeditions, women sold kola nuts and purchased raw cotton for spinning. When they arrived in places like Jenné, women sold their thread for cowrie shells, the regional currency, which they used to buy salt. Jula men exchanged cloth for salt, kola, and cowries along trade routes.[6]

Beyond these descriptions of rural economic activities along major trade routes, the reader has little idea about agricultural systems within the communities Callié visited. He fails to note, for example, the nature of farming techniques and the organization of agricultural work. He does, however, comment on the prevalence of slave labor. Among the Malinké in particular, he notes that relatively prosperous households depended upon slave labor in agriculture. Poorer, non-slave-owning households were obliged to cultivate their own fields. Slaves also served as porters in long distance trade caravans. When the caravan stopped in a village at the end of the day, slaves immediately went to the bush to collect wood for preparing the evening meal.

Tengréla was an important trade center when Caillié passed through in January of 1828. It was the converging point of caravan routes from the Guinea Coast and Niger River, where traders exchanged salt for cloth and kola nuts.[7] The entrepôt also contained a thriving indigenous handicraft textile industry whose cloth was traded as far north as Ségou and Djenné.[8] Tolls were levied along the major caravan routes on the quantity of merchandise transported. Customs agents bearing whips and wearing black-hooded robes adorned with ostrich feathers stood at the outskirts of villages collecting tolls.[9] This revenue made the chiefs of well-placed villages relatively rich individuals.

Sixty years later, Captain Louis-Gustave Binger appeared in the Tengréla and Niellé areas of northern Côte d'Ivoire during his famous mission of exploration and reconnaissance for the French government. He left Bamako in June 1887 and traveled through Tiéba's empire of Kénédugu at the time when Samori's troops were unsuccessfully laying siege to the empire's political center at Sikasso (Plate 1). Binger hoped to visit Tengréla to display French wares, to engage in treaty making, and to find out more about the country to the south. However, he was denied entry to the important trade center because of France's friendly relations with Samori (Plate 2), who had seized Tengréla in 1885 only to be driven out a few months later. Tengréla was allied with Tiéba in the conflict and thus viewed the French with great hostility. In fact, Binger had to flee Tengréla in the middle of the night at the risk of being slain.[10]

During his travels through the Niellé chiefdom, Binger made a number of interesting observations on agricultural production, trade, and settlement patterns. Like Caillié, Binger traveled through northern Côte d'Ivoire during the mid-dry season months of January–February. Thus, most agricultural activities had ceased, and Binger's observations were limited to late cotton harvests and the mounding of yam fields. His comments on agricultural activities in the chiefdom of Niellé provide a glimpse into the organization of production. For example, Binger noted that Pegué, the chief of Niellé, settled his slaves in field camps encompassing up to fifty individuals. They farmed and stocked produce in granaries to which Pegué had free access. Stopping in the village of Dioumanténé about 100 kilometers north of Katiali, Binger noted:

> The Mandé-Jula who are settled here have their slaves weave white bands of cotton streaked with blue similar to that of Fourou; in the village where I camped there were seventeen looms in operation. Cotton is harvested here; there are fields everywhere but I didn't see any indigo; the village also grows maize and different varieties of millet and sorghum, some peanuts and a lot of rice.[11]

Outside of Niellé, Binger stopped at an encampment where slaves were watching over the recent harvest. Yams, millet, sorghum, and cotton were the principal crops. He noted that slaves played an important part in the cultivation and spinning of cotton. "Every morning" cotton was harvested by slave women and spun into fiber at night.[12] Denied entry into Niellé by Pegué, Binger moved on to the eastern frontier of the chiefdom. In the village of Léra, Binger recorded seeing 600 kilograms of cotton ("31 foufou de 20 kg") in the market.[13] He reported that Jula traders bought raw cotton for both free and slave women to spin. The finished cloth, woven by Jula men, was exchanged for salt "and other rich products."[14] Neither Caillié nor Binger inform us on how cotton was grown or the varieties cultivated. We need to turn to other sources for information on these features of late-pre-colonial cotton growing. What type of cotton was cultivated and traded in the Ivorian savanna at this time?

African cotton

The origins and evolution of cotton are intriguing because the plant was domesticated independently in four different parts of the world.[15] The genus *Gossypium* contains four distinct lint-bearing species: two are from the Old World (*G. arboreum* and *G. herbaceum*) and two from the New World (*G. barbadense* and *G. hirsutum*). On the basis of chromosome size and structure, botanists believe that a wild A-genome diploid was the ancestor of all linted cottons, and that it came from Africa. One of its descendants, if not the ancestor itself, is *G. herbaceum* subsp. *africanum*, which grows wild in the drier savannas of Zimbabwe, Botswana, and Mozambique.[16] It is the only wild species of *Gossypium* that bears lint.[17] The Old World domesticated cottons, *G. herbaceum* and *G. arboreum*, both developed from *africanum* in the Middle East (southern Arabia) and South Asia (India and Pakistan). These so-called "Asiatics" re-invaded Africa and spread across the savanna belt long before the rise of ancient Egypt. In contrast, the native cottons of the New World were lintless. It was only after *Gossypium herbaceum* var. *africanum* (or one of its ancestors) combined with wild types that the New World linted species *G. barbadense* and *G. hirsutum* developed.[18] Varieties of both the Old and New World species were under cultivation in the West African savanna at the end of the nineteenth century when the French botanist Auguste Chevalier recorded their presence.

Chevalier encountered what he believed were four distinct cotton species during a scientific expedition to French West Africa in 1898–99 organized by General L. E. Trentinian, the Lieutenant-Governor of French Soudan.[19] These species included (1) *G. herbaceum L.*, which he called the "oldest species known by the natives;" (2) *G. barbadense L.*, var. *Sea Island*, introduced into the Soudan in 1896 "thanks to district commanders who distributed seeds among chiefs of the principal villages"; (3) *G. religiosum L.*, "a hardy species that one sometimes finds near the coast of Senegal"; and (4) *G. punctatum Perr.*, "a very hardy African race of *G. hirsutum L*. . . that is found in virtually every village."[20] Hutchinson, *et al.*, classify *G. religiosum L.* as the *punctatum* variety of *G. hirsutum*.[21] Thus, if the last two species are one and the same (*G. hirsutum* var. *punctatum*), then Chevalier only encountered three species: one from the Old World (*G. herbaceum*) and two from the New World (*G. barbadense* and *G. hirsutum*).

The archaeological record shows that *G. barbadense* had evolved into a coarse, short staple cotton in western South America by 4500 BP. Over the millennia, it continued to differentiate into a number of varieties, the most famous being the long-stapled and fine-textured perennial known as "Sea Island." Sea Island evolved in the West Indies and the tidewater areas of Georgia and South Carolina, where it was established as a perennial by mid-eighteenth century. An annual form rapidly evolved, most likely from crossing with a wild form of *G. hirsutum*,[22] but it was the perennial form of *G. barbadense* that spread to West Africa and eventually to Egypt, from eastern South America and the Caribbean.[23]

G. hirsutum is believed to have evolved in southern Mexico and Guatemala. Two *hirsutum* varieties, *punctatum* and *latifolium*, spread widely outside this core area. As Chevalier noted, the *punctatum* variety became the most commonly grown cotton in West Africa by the end of the nineteenth century. One of the reasons for its popularity may have been the relative ease with which it could be hand-ginned.[24] However, in Côte d'Ivoire it was the *latifolium* variety that was developed into commercially successful annual varieties like *Allen* during the post-World War II period. *Latifolium* originated in the Chiapas area of Mexico. Annual forms of this perennial were selected in the southern United States, which resulted in the rise of the world famous Upland cottons. These medium staple and highly productive strains spread rapidly to the major cotton growing regions of the world during the nineteenth and twentieth centuries. Colonial agronomists conducted field trials at experiment stations and distributed "improved" varieties of *barbadense* and *hirsutum* to peasants during the era of forced cotton (chapters 3 and 4).

Botanists believe that the New World varieties first entered Africa between the sixteenth and early eighteenth centuries in the course of the slave trade.[25] Genetically speaking, these "introduced" cottons were half African in origin, since all of the tetraploid cottons of the New World contain African A-genome diploids.[26] They were also subject to considerable experimentation and selection by African cotton growers. For example, it is only in West Africa that an annual form of *punctatum* is found. Moreover, West African *punctatum* is the only New World cotton that is highly resistant to bacterial blight disease.[27] When colonial agronomists first began studying local varieties in detail, they were struck by the high degree of hybridization of African cotton. The head of the Côte d'Ivoire Agriculture Service indicated in 1915 that hybridization was a continual process as farmers acclimated newly acquired varieties to suit local soil and climatic conditions. He noted that a wide range of hybrids could be found growing in the same field.[28]

Chevalier described the cultivation of perennial cotton in the middle Niger River valley. In the Djenné area it was generally sown during the early part of the rainy season (June–November). Ten to twenty seeds were placed in holes spaced 70 to 150 cm apart. Once the plants had emerged, they were thinned so that just 2 to 3 plants remained in each place. Cotton was typically intercropped with sorghum in the Djenné region. After the sorghum harvest, the soil was mounded around the young cotton shrubs. In some locations, the plants were watered during the dry season and compost originating from refuse piles at the edge of villages was added to the mounds to improve soil fertility. If sown early enough in the rainy season, the shrubs would yield some cotton during the first year. Yields were highest during the second and third years after plants had been cut back to the trunk to produce a dense growth of new shoots. Some plants were not pruned and continued to produce cotton throughout the year, albeit of inferior quality. After 4 to 5 years of continuous cultivation, yields declined and farmers planted leguminous crops like peanuts and ground peas in place of cotton. Although Chevalier does not comment on it, perennial cotton was also less labor demanding than annual varieties which needed to be planted every year.[29] Further south in the humid savannas of the French Soudan and northern Côte d'Ivoire, perennial cotton was intercropped with maize, millet, yams, peanuts, rice, and various condiments (see chapter 3).

This brief overview of West African cotton varieties raises three general points. First, in contrast to the dominant cotton development narrative that emphasizes the critical role played by Europeans in introducing new cotton varieties to West Africa, these "exotic" varieties were in large part derived from African sources. Half of the genetic complement of New World cotton varieties is of African origin. Second, African cotton growers actively experimented with new varieties in their fields. They favored plants that produced good yields and possessed qualities demanded by the local handicraft industry (color, fiber length, ease of hand-ginning). This image of African cotton growers as innovative agents contrasts with the conventional representation of them as passive recipients of introduced technologies. Third, in light of the pre-colonial regional handicraft textile industry observed by Caillié and Binger, African men and women had economic as well as agronomic interests in new cotton seeds. As Richard Roberts shows in the case of colonial French Soudan, female cotton spinners were quick to appreciate the labor-saving qualities of medium and long-staple cotton. Hand carders and manual cotton gins introduced by the Colonial Cotton Association (Association Cotonnière Coloniale) at the turn of the century were also highly valued for improving labor productivity in the handicraft industry.[30]

In short, cotton growing, spinning, and weaving were important elements of African rural economies and farming systems long before European explorers and colonists began to contemplate their potential for serving the Metropolitan textile industry. The social organization of production based on domestic slavery, the inter-regional trade in cotton textiles, and the dominance of perennial varieties in intercropped fields were soon to be radically altered in the context of European colonialization. Military conflicts among African and European empire builders in the last two decades of the nineteenth century set the stage for the ensuing agrarian transformation.

The Korhogo region on the eve of colonization

As Tiéba expanded the boundaries of his empire to the south and east, Samori's forces moved into the Bagoué River valley on the western frontiers of Kénédugu. Samori was being squeezed to the southeast as French military forces expanded into the upper Niger River valley in the early 1880s.[31] The competition for territory between Samori and Tiéba erupted into the siege of Sikasso by Samori in 1887–88. During this sixteen-month conflict, Korhogo and its "vassals" (including Katiali) supplied vital food supplies to Sikasso.[32] The provisioning of Samori's 10,000–20,000 troops was much more difficult. Binger reported that approximately 200 porters arrived *each day* on the battlefront carrying about two tons of food grains for Samori's hungry troops, and the source area of this food was more than 300 kilometers from the front. As the area between the Baoulé and Bagoué rivers had been devastated since 1885, food staples and porters had to be requisitioned beyond this depopulated zone.

The level of tribute exacted by Samori and Kénédugu varied from region to region and village to village. Some villages, usually the longtime supporters of Samori or Kénédugu, were exempted from the predations and requisitions of their soldiers. Villages inhabited by captives appear to have paid a larger "tribute." Others, usually those that were forcibly conquered, had to pay the highest tribute.[33] In both empires, the level of requisitions was not only uneven but also irregular over time. When Samori lay siege to Sikasso in the late 1880s, tribute seems to have been relatively high and regular on both sides. In other periods, the level and timing of tribute varied according to an area's location vis-à-vis a battlefront, the history of its political relations with Samori or Kénédugu, and the policies of each empire's representatives in that area.[34] Whether requisitions and pillaging were frequent or periodic, the heavy reliance of both empires on food and slaves left the countryside in ruins. Binger observed ruins everywhere in the Niellé region (also known as the "Follona" of Pegué) where troops from Kénédugu regularly raided villages (Plate 3). As a result of enslavement, famine, and war-related deaths, Binger estimated population densities to have declined in the Niellé area from an estimated 40 inhabitants per square kilometer to 12 to 15 persons per square kilometer by the late 1880s.[35]

In January–February 1894, Jean-Baptiste Marchand, a French explorer and military officer, traveled through the Koroko (Korhogo) region during a reconnaissance mission. His objective was to determine a navigable route linking the Niger river system with the Guinea coast. He decided that the best route of what was optimistically named *Le Transnigérien*, would link the Bandama and Bagoué rivers. However, this would require building a 95 kilometer overland railroad to join the two river systems, and the proposed railway would have passed just south of Katiali.[36] Marchand spent the night in M'Bengué and passed through Katiali where he regained "the caravan route" linking Tingréla and Kadioha. His official report is scant on most details of his expedition in Senufo country.[37] What the report does convey is a sense of the devastation wrought by Kénédugu under Babemba in 1893, when he pillaged and razed much of Senufo country.

> [T]he spectacle of horror that incessantly unfolded before our eyes during this long journey did little to comfort us; fields laid waste, villages destroyed and still smoking from which rose the veritable odor of a charnel house – at the outskirts, under the rare trees offering some shade where one was tempted to seek shelter against the excesses of a torrid and stinking heat, [we witnessed] a few people with smallpox, almost always elders, on the verge of death and moaning their last lament over the bodies of those who were already dead . . .[38]

In summary, the impressionistic views of rural life provided by Caillié, Binger, and Marchand indicate that dramatic changes occurred in northern Côte d'Ivoire between the first and last quarter of the nineteenth century. These changes include the political and economic incorporation of small chiefdoms into the expanding empires of Kénédugu and Samori. This process involved tremendous instability, dislocation, and depopulation. The incessant raiding and slaving also appears to have transformed the organization of production in the countryside. Slave labor in agriculture appears to be much more widespread in the 1880s than in the 1820s. We also witness the institution of forced cultivation in the faama's and almani's fields as a mechanism that enabled the expansion of these African empires.

At the time of Marchand's passage in January 1894, Babemba had already withdrawn to Sikasso to deal with a revolt among the Miniankala (Minyanka). Some of his troops stayed behind to construct outposts in newly conquered villages like Tiému and Tioroniaradugu. They quickly evacuated the area, however, when they learned that Samori's troops were approaching from the northwest. Samori was being pushed out of

Plate 1 (L–R) Babemba Traoré and Tiéba Traoré, and unidentified military chief, Sikasso (Mali) (*c*. 1892)

Plate 2 Samori Touré, holding the Koran, St. Louis (Senegal) (*c.* 1898)

Plate 3 Ruins of old Niellé, 15 January 1888

Plate 4 Gbon Coulibaly, Canton chief of Korhogo (1943)

Plate 5 Samori listening to Lieutenant-Governor Trentinian's speech following his defeat, Kayes, 1898

Plate 6 "Burned houses, cadavers, and the dying after Samori passed through the village" (27 September 1898)

Ouassoulou (Wasulu) by the French military forces commanded by Lieutenant Colonel Etienne Bonnier, who spearheaded the advance into the Buguni area.[39] As Samori moved east of the Bagoué, he took control of areas that were previously loyal to Kénédugu, such as Tengréla and Kaloa.

Samori settled in the village of Diégbé, five kilometers to the north of Korhogo, where he called a meeting of all the chiefs of the region. Gbon Coulibaly was at the head of this delegation (Plate 4).[40] The chief of the Kiembara subgroup shrewdly submitted to Samori's authority by declaring: "I am not a warrior but a cultivator. I do not want war. I entrust myself to you."[41] Samori accepted Gbon's pledge of allegiance and proceeded to seize grain and slaves to support his expansionist policy. The "Diégbé pact" was to have far reaching consequences for the spatial distribution of the region's population. Spared from the worst of Samori's war machine, the Korhogo region became a safe haven for refugees fleeing the ravages of both Babemba's and Samori's troops. As a result, the area became increasingly populated while communities in the hinterland either lost their population or regrouped into larger agglomerations for defensive purposes.[42]

Plate 7 View of Korhogo, *c*. 1943; the *jasa* residential quarter is located in background to right (*c*. 1943)

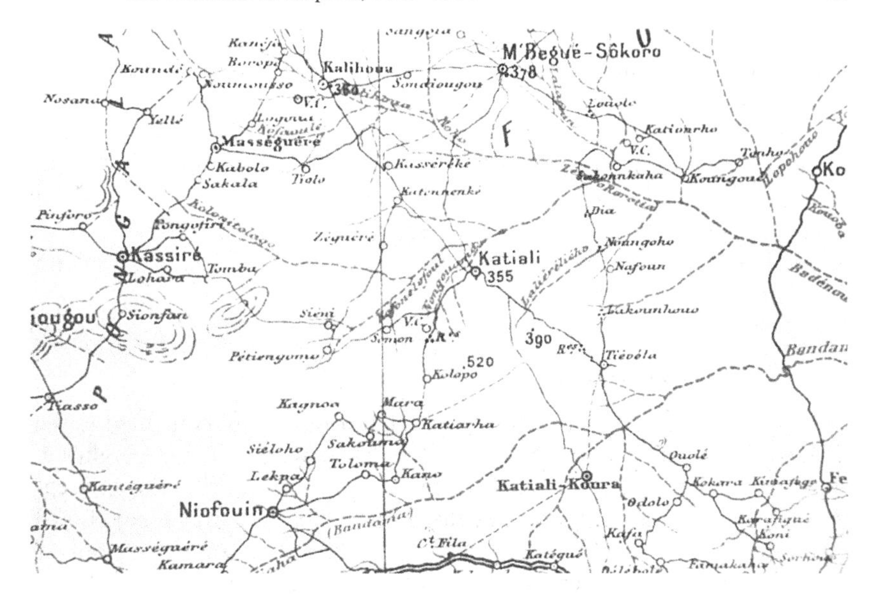

Fig. 2.2 Map showing the location of Katiali-Koura and Katiali (*Source:* ANSOM AF 1863; "Korhogo," Feuille No. 1, Service Géographique des Colonies, by A. Meunier, 1:500,000 [1923])

According to Yves Person, Samori hoped that Kénédugu would serve as a buffer between his new eastern empire and the French, who were concentrated along the Niger River. To avoid any further conflicts with Kénédugu, Samori established a buffer zone between the southern reach of Kénédugu and the northern Korhogo region.[43] In January 1895 his *sofas* forced the population of the M'Bengué chiefdom to evacuate the region and resettle to the west of Korhogo at what is today the village of M'Benguébugu.[44] The people of Katiali moved to an area south of the Bandama River and created the village of Daluru ("Five Gateways"), which appears on a late nineteenth-century map as Katiali-Koura ("New Katiali") (Fig. 2.2). A garrison was left behind in Kaloa to watch over the deserted area. This "tacit peace" was formalized in August of 1896, when Babemba and Samori had an official reconciliation in which territory and hostages were exchanged and trade relations were opened.[45]

The improved relations between Babemba and Samori greatly concerned the French military, who feared a united front against their advance into the southern savannas. As late as June 1897, the French had hoped that Kénédugu would help them defeat Samori.[46] The primary goal of the French remained linking their possessions along the middle Niger with those along the Guinea coast.[47]

Between May of 1896 and May of 1898, Samori held a firm grip on the Korhogo region. A military government backed by an army of 2,000–3,000 sofas ruled by Bilali in Féhéré (Fègèrè) insured the flow of foodstuffs to those parts of the empire that had difficulty in provisioning itself. The food for peace arrangement ultimately brought terrible hardships to Senufo communities during these last two years of Samori's rule. The heavy and systematic requisitions were bitterly resented by Senufo farmers.[48] Famine was not uncommon. There are reports of village chiefs foraging for food and dying from eating poisonous wild yams.[49]

Redrawing the map

The French conquest of Kénédugu and Samori is well documented in the annals of French military history.[50] After two weeks of fierce fighting, superior French firepower breached the walls that Samori found impregnable. Babemba committed suicide rather than submit to the French. Local influential chiefs who agitated against the French prior to the fall of Sikasso were "severely repressed".[51] When French troops descended into the Korhogo region in May of 1898, Samori's forces quickly fled south to join the almani south of Touba. Sensing a dramatic shift in power relations, Gbon Coulibaly took the lead once again and rallied local support behind the French. Famine was raging in the region after food stores were plundered by Samori's fleeing *sofas*. The French quickly occupied the deserted region, establishing posts at Tombugu, Tiému, and Boribana. Samori's retreat ended on 29 September 1898 when he was captured by French forces near the Liberian border (Plate 5).[52]

With the fall of Kénédugu and Samori, the French wasted little time in occupying the northern savanna region of Côte d'Ivoire. The military's ability to occupy the area was undoubtedly facilitated by the physical dislocation and insecurity reigning in the region. In the wake of the successive massacres, slave raids, and pillaging by Babemba's and Samori's armies, the French realized that it was an opportune time to occupy the north (Plate 6). As the Lieutenant Governor of the Soudan (Mali) wrote in a telegram to the Inspector General of the Colonies in July of 1898:

> It is urgent to profit from people's stupor and terror to occupy securely [the] entire country in wake of invasions; unfortunately, we do not have enough soldiers . . . if we do not immediately occupy, we risk being forced to reconquer bit by bit during the next campaign when terror will be dissipated.[53]

The first steps taken by the French were to establish an administrative organization and to impose a head tax on the local population. Both tasks were accomplished relatively quickly by military officials. In 1898 the Korhogo region became part of an administrative area called the Upper

Bandama Circle, which also included the districts of Tombugu and Odienné. The Upper Bandama, Bouna, and Bouaké Circles together comprised the Military Territory of Kong. This territory was dependent on the Volta Region and attached to the French Soudan; it was not until April 1900 that the Korhogo region was incorporated into the colony of Côte d'Ivoire.[54] Figure 2.3 illustrates the new political geography under French colonial rule. In contrast to Fig. 2.2, this new map of the French Soudan represents a dramatic reorganization of power relations in the region. A copy of the map was sent by Lieutenant Governor Trentinian of the French Soudan to the Minister of the Colonies. In a letter accompanying the map, Lieutenant Governor Trentinian wrote:

> From the point of view of the political organization of the Soudan territory, it is divided into regions, variously colored, themselves subdivided into districts (*cercles*), whose limits are indicated on the map. Finally, conventional signs allow one to distinguish territories administered directly from those under our immediate protection, and a legend indicates the population of each circle and the land area of each region.[55]

Lieutenant Governor Trentinian's and the cartographer's assertions to territorial control were, however, fiercely contested by indigenous peoples. For example, the Gourou and Baoulé resistance to French colonial rule in the Zuénoula and Bouaké regions effectively countered the claims being made in such maps.[56]

Head taxes were swiftly imposed upon the population in the newly occupied territories. Just two months after Samori's capture, the Minister of Colonies instructed the Lieutenant Governor of the French Soudan "to take advantage of the presence of our troops to institute immediately the collection of a head tax such as it already exists in other parts of the colony." It was important, he noted, that people become accustomed to paying such a tax, so that when a civilian administrator assumed control of the area it would already be in force.

Redrawing the map also involved an accurate recording of the population distribution in the newly conquered territory. The successive raids into the Korhogo region by Babemba and Samori resulted in major changes in settlement patterns. The dispersed pattern of small villages once characteristic of the chiefdom of Katiali had changed to a pattern of either large village-fortresses or regrouping in protected zones, like the increasingly densely populated region surrounding Korhogo. Maps such as the one shown in Fig. 2.2 show the location of the refugee village of Katiali-Koura and the abandoned village of Katiali. The ruins (identified as "R^{es}" on the map) of former villages appear to the south and southeast of Katiali. In early 1900, with the permission of the French military commander, the inhabitants of Katiali-Koura returned to Katiali.[57] Some of the refugees

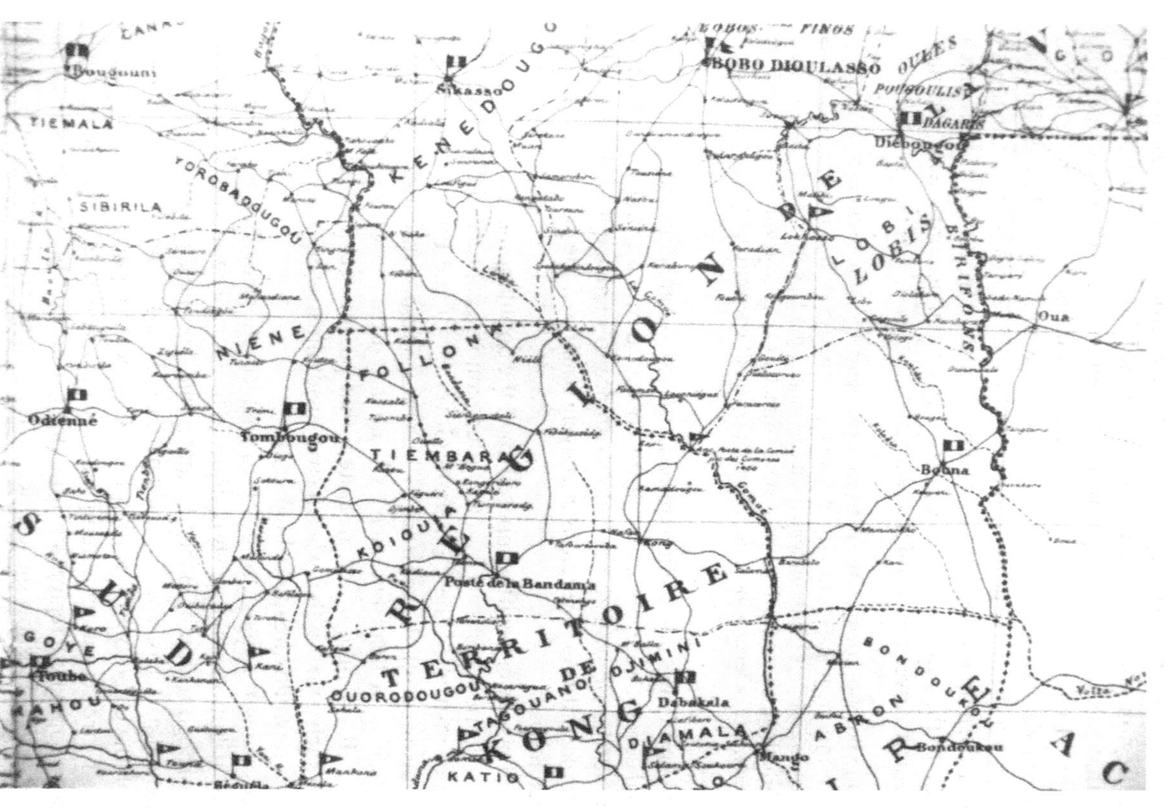

Fig. 2.3 Map showing colonial political divisions (*Source:* ANSOM S.G. Soudan, dossier 9, "Carte politique du Soudan Français," *c.* 1899)

from former villages within the chiefdom decided to settle in Katiali itself, where they established new residential quarters.[58] Most decided to stay in the areas where they had sought refuge.

Research by the Ivorian geographer Sinali Coulibaly shows that the demographic impact of the late-nineteenth-century turmoil was still evident in the population distribution recorded in the 1975 census. His analysis indicates that the size of settlements tends to be much larger in the areas east and west of Korhogo. In these areas, 60–80% of the population live in villages of greater than 700 inhabitants. Large, uninhabited areas are commonly found in between these communities. In contrast, villages in the subprefecture of Korhogo are more numerous, dispersed, and much smaller in size. The population density map in Fig. 2.4 shows the highly uneven distribution of the region's population in 1975.

The demographic legacy of the pre-colonial period was a continued source of frustration to colonial administrators who wished to extend their authority throughout the region. As late as 1958, the Korhogo Circle Commander exclaimed that it was "impossible to govern correctly" due to the uneven distribution of the region's population. Commander D. Pinelli noted that in the Korhogo subdivision there were 200,000 persons divided into 14 cantons and 748 villages distributed over an area encompassing 11,000 km^2.[59] It was also difficult and rare for administrators to visit villages outside of this relatively densely settled zone. When the commander of the Korhogo circle visited Katiali in 1942, it appeared to be the first such visit in many years.[60]

Confronted by the logistical difficulties linked to the region's population geography, colonial officials organized an administrative structure that allowed them to intervene more effectively in local political and economic affairs. The chain of command and surveillance was hierarchical in its administrative and spatial forms. Within each administrative district (*cercle*), the chain of command began in the office of the district officer or commander (*Commandant de Cercle*). For example, if it was time to announce the area to be cultivated in cotton, he would first inform the French administrators heading each subdivision in the district. Each subdivision was further divided into cantons at the head of which was an indigenous chief who transmitted the district officer's directives to village chiefs within his canton. Village chiefs, in turn, informed the lineage heads of residential quarters about the cotton area to be cultivated that season. To facilitate the flow of information and orders, each canton was required to send a representative to the administrative center of each subdivision where he lived in a special residential quarter called the *jasa* (Jula: encampment) (Plate 7). The representative was known as the *jasatigi* (Jula: jasa representative). In the mid-1920s, the Korhogo district (*Cercle de Korhogo*) included

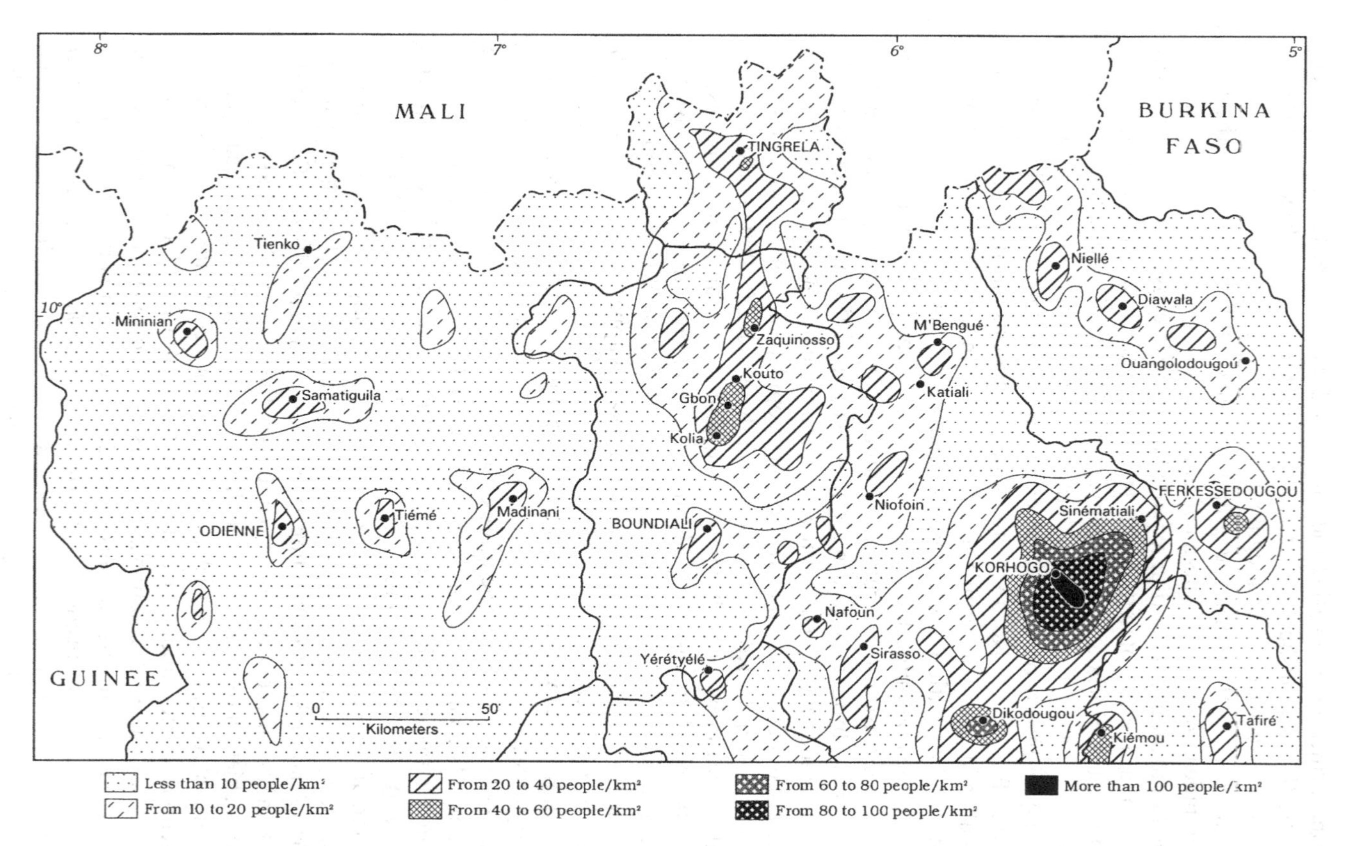

Fig. 2.4 Population densities of the Korhogo region, *c.* 1975 (derived from Coulibaly, *Etat*, p. 53)

the Korhogo, Boundiali and Ferkéssédugu subdivisions; there were twenty-five jasatigis in the Korhogo subdivision.[61] It was the jasatigi who informed the canton chief of the district commander's orders. The district commander did not speak directly to the jasatigis. There were two additional intermediaries: the commander's interpreter, and the head of the jasatigis, known as the *jasakuntigi* (Jula: head of the jasa representatives).[62] The district commander and his subdivision chiefs employed a local police force (*garde de police*) to see that their directives were in fact carried out. These "guards", as they were locally known, were the most visible and brutal representatives of the colonial state at the village level.[63] In 1932 there were thirty-eight guards assigned to the Korhogo subdivision, eighteen to Ferké and another eighteen to the Boundiali subdivision.[64] When circumstances required it, district guards recruited other Africans to assist them. These irregulars were locally known as "goumiers."[65] Examples of this command system in action are provided in the following chapter. My objective here is to locate the origins of this colonial administrative structure in the context of the political upheaval and population redistribution of the late pre-colonial period.

Conclusion

During the last two decades of the nineteenth century, the Korhogo region was at the center of a violent episode in empire building and predation by African and European powers. The value of Binger and Marchand's accounts is that they provide firsthand reports of the geographical location, expansion, and shifts of African empires on the eve of French colonial rule. Their emphasis on devastation, displacement, and carnage is somewhat colored by their own ideological missions of peaceful trade and colonization, also known as the "mission civilatrice". Nevertheless, "the spectacle of horror" which emerges from these late pre-colonial accounts informs us about the great social and political upheaval pervading the region at this time. These highly unstable conditions contrasted with the relatively peaceful period reported by Caillié.

The great instability of the late nineteenth century ultimately facilitated French occupation of the area. The *Pax Gallica* was a welcome change to the Senufo and Jula inhabitants of the region. However, the continuities with the past were not so well received. Within the first decade of colonization, the population was to experience the French version of food requisitions and forced labor. Once again, the region was to become a granary and the Kiembara the dominant ethnic group. The political fortunes of Gbon Coulibaly, the premier collaborator, were to rise throughout the colonial period and well after independence. As in the past, resistance to foreign rulers and their collaborators was to continue as well. Indeed, the themes of appropriation, collaboration, and resistance are major bridges spanning the precolonial and colonial periods in the social history of the Korhogo region. In the following chapter, I pick up the thread of the indigenous cotton growing and handicraft industry to examine how local social and economic processes influenced the pace and direction of agricultural change in northern Côte d'Ivoire up until the end of the Second World War.

3 The uncaptured corvée, 1912–1946

> It seems paradoxical that our spinning mills of northern Normandy and Alsace can be lacking in primary materials when cotton plants grow spontaneously in many regions of our colonial Empire . . .
>
> *La Dépêche*, 14 Novembre 1928

The failure of rural producers to market sufficient quantities of cotton fiber for France's textile industry is a recurring theme in the history of cotton in colonial Côte d'Ivoire. Repeated attempts by colonial administrators to intensify cotton cultivation were never realized in a sustained and satisfactory manner. France's African colonies furnished just one to three percent of metropolitan cotton fiber needs over nearly three and a half decades.[1] Depending on the period, officials cited a number of factors to explain the disappointing results: low-yielding varieties, cotton pests, low prices, food crop competition, and peasant resistance to forced cultivation. A particularly frustrating factor for administrators was the existence of a local cotton market that supplied the indigenous weaving industry. This parallel market represented a direct challenge to their stated goals of furnishing the metropolitan textile industry with its basic raw materials and exporting cloth to the colonies. How to deal with this "problem" of local cotton consumption was hotly debated among colonial officials. Some sought to suppress the indigenous handicraft industry, while others condoned it. There was some consensus that in order for cotton exports to increase, local needs had to be met first. As long as free trade principles were adhered to, this meant that cotton output had to rise substantially if both local and export markets were to be satisfied. The aim of cotton policies was thus to encourage producers to grow more cotton than could be used locally. The record shows that this aim was achieved but rarely to levels that were deemed satisfactory. Colonial administrators repeatedly resorted to force in order to capture a larger share of peasant cotton. However, disputes among administrators over the wisdom, morality, and legality of compulsory cotton resulted in its uneven enforcement and shifting policies. Cotton production followed a boom-bust pattern (Fig. 3.1) mirroring swings in administrative

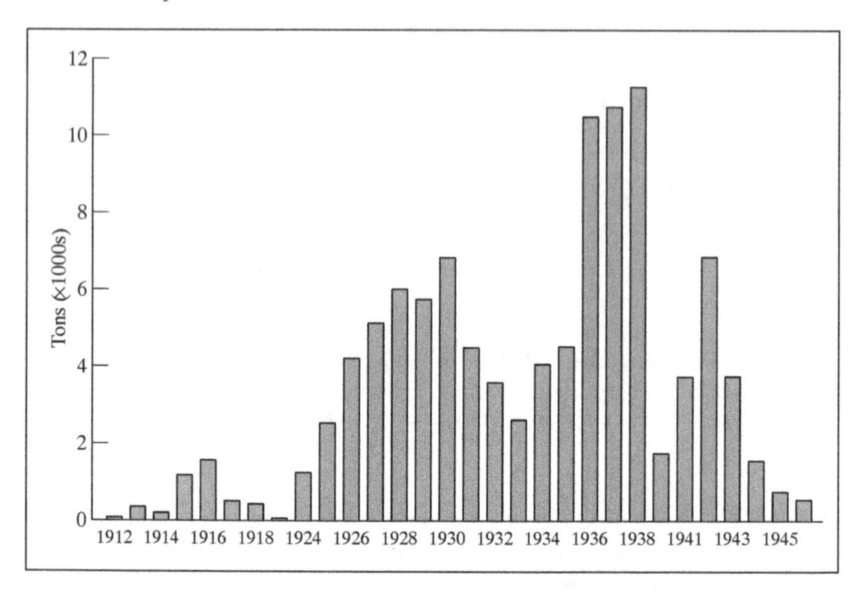

Fig. 3.1 Côte d'Ivoire cotton production, 1912–46 (*Source:* ANRCI)

coercion, policy reforms, and peasant resistance to cotton policies. The three peaks in cotton exports (1930, 1938, 1942) correspond to periods of heightened coercion and hardship in the cotton growing areas (Fig. 3.2). This chapter argues that the parallel cotton market not only offered an alternative outlet to cotton producers but also played a part in shaping the nature of colonial agricultural policies.

The persistence of the parallel cotton market and the influence it exerted on colonial policies illuminate a number of issues in African agrarian social history. First, its very existence corrects the erroneous view that cheap textile imports were the death knell of indigenous handicraft industries in Côte d'Ivoire.[2] Imported French cloth was rarely cheap; indeed, its high price was often a stimulus to local weaving. Second, and of more theoretical importance, this history shows that France's failure to capture peasant cotton was in large part due to the relative freedom enjoyed by rural producers in cotton markets. The colonial state's general adherence to a free trade policy allowed local Jula traders supplying the indigenous handicraft weaving industry to compete with French merchants over cotton supplies. Local traders invariably paid higher prices and offered easier marketing arrangements than European merchant houses. Cotton growers took advantage of this parallel market when it was in their best interests. Much to the chagrin of colonial administrators, extension agents, and French

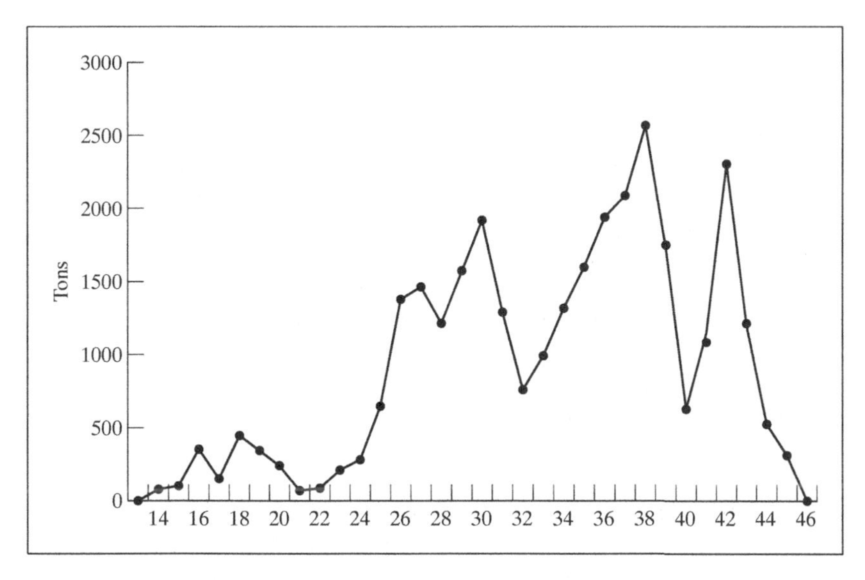

Fig. 3.2 Cotton lint exports from Côte d'Ivoire, 1913–47 (*Source:* ANRCI)

textile firm representatives, the staple food trade also offered producers an alternative outlet for their labor. This second market option encouraged peasants to neglect cotton in favor of yams or rice when the prices of these foodstuffs were more attractive than cotton. In the end, the relative autonomy of peasants to take advantage of these market opportunities effectively thwarted colonial attempts to expand cotton exports.

The freedom of peasants to sell their crops to whomever they wanted was linked to how local administrators interpreted and the extent to which they enforced the free trade policy. Their interpretation was influenced in part by their personal ambition to advance their careers by meeting production and export targets set by their superiors. Policy implementation was also influenced by debates taking place in administrative districts about how best to achieve the goal of provisioning France's textile industry with its basic primary material. Should cotton be grown as a monocrop in collective fields or as an intercrop in individual fields? Should payments be made to village chiefs or to the individuals who actually produced the crop? Should peasants be forced to grow and market cotton beyond local needs? Should local weaving be suppressed altogether to stimulate demand for imported French cloth, or should it be encouraged to stimulate cotton production? Should savanna peoples be encouraged to migrate to the tropical forest region to work on European plantations or be allowed to stay at home and cultivate cotton? As Roberts and Isaacman demonstrate in their case studies of cotton colonialism in the French Soudan and Mozambique, the questions and dilemmas centered on cotton were simultaneously questions about the function and role of the colonial state, about the relationship between colonial authorities and their subjects, and about the limits of colonialism as a political-economic strategy of European imperialist powers. For the purposes of this discussion, such debates over appropriate economic and agricultural policies are important because of their influence on the latitude peasants had in selling their cotton surpluses in local markets. The principle of free trade and its enforcement were critical to peasant autonomy. As this chapter shows, its interpretation and implementation were influenced by particular conjunctures (world wars, economic downturns), the ambitions of administrators at all levels, the commercial practices of African and European merchants, criticisms by metropolitan-based colonial authorities, and resistance by local populations.[3]

In general, administrators realized that to increase output and to satisfy the metropolitan textile industry's raw material needs, three conditions had to be met: (i) export market prices had to be competitive with local market prices; (ii) cotton area and yields had to increase and farmers had to devote more of their labor to cotton; and (iii) the quality of local cotton had to be attractive to French spinners and weavers who would only purchase fiber that was of the desirable color, length, and strength, as well as price. As Roberts shows in the case of the neighboring French Soudan, French colonial administrators, agricultural advisors, and textile company representatives alike viewed the problem of increasing cotton exports as a technical problem that could be solved by using different tools. He conceptualizes a tool kit with three compartments (cotton production (labor), marketing, and quality) which were the realms in which policy makers intervened to promote export-oriented cotton production. The idea that cotton development could be reduced to a series of technical interventions disguised the politics of development and especially the disagreements among administrators, agronomists, and merchants about which mix of policies to promote. Roberts's observations on colonial cotton policy ("la politique cotonnière") and development discourse in Soudan echo the case of Côte d'Ivoire during this period.

> The debates among the contending groups of colonial officials, entrepreneurs, and Africans were not merely about the most efficacious means of encouraging cotton production and controlling the cotton harvest of the Soudan. These debates must be seen as part of a colonial development discourse. As such, cotton colonialism also produced cultural representations of Africans, of their societies and economies, of their capacities to conform to prevailing European cognitive categories of laborers and peasants, and the roles and functions of colonialism.[4]

This chapter suggests that the colonial development discourse was neither hegemonic nor successful in achieving its stated goals. Disagreements among administrators, merchants, and agronomists on how best to encourage farmers to increase output resulted in opposing discourses and contradictory policies. A close reading of archival documents reveals at least three distinct development discourses. The *rational peasant development discourse* was based on the idea that peasants would produce cotton for export markets if provided with the right economic incentives. It viewed farmers as knowledgeable and rational economic agents who would increase cotton supplies if purchase prices were sufficiently attractive and if they were allowed to grow cotton following customary methods. It emphasized competitive markets, payments to individual producers, building upon indigenous farming practices such as intercropping cotton in food crop fields, respect for local knowledge, and an evolutionary approach to technological change. The role of the colonial state was to ensure that markets operated freely, that prices were competitive, and that the development and extension of high-yielding cotton varieties was supported. The *compulsory development discourse* viewed Africans as unwilling to produce cotton for export markets unless forced to do so by the colonial state. This generally racist view represented Africans as apathetic and lazy by nature, and unwilling to produce cotton beyond local needs. They would resist working hard for monetary gain unless forced to see it was in their best interests to do so. It emphasized production quotas, collective fields, monocropping, forced deliveries, quick action, and punishment for failing to follow the instructions of insistent administrators. The colonial state's primary role was surveillance – to make sure that peasants were following orders and to punish those who failed to deliver. An intermediate *paternalistic development discourse* combined elements of these first two. Depending on the circumstances, it encouraged both collective and individual fields and payments to individual producers, and combined compulsory cultivation with free markets. The role of the colonial state was to impose cotton growing yet to encourage market competition, to promote research and development, to defend local populations from the excesses of certain administrators and merchants, and to conduct surveillance.

The coexistence of these different visions and modalities of economic development resulted in a diversity of situations in which Senufo farmers and Jula traders participated in the colonial economy. Disputes between different agents of the colonial state, private sector, and local populations on how to intervene in the rural economy gave African peasants and merchants a measure of freedom. This relative autonomy is best expressed in the remarkable resilience of the handicraft industry to colonial depredations. The local cotton market supplying this industry provided an outlet for peasants and traders alike that colonial administrators found difficult to control.

At the same time, the intrusion of the state and expatriate firms in the daily economic and social life of Senufo and Jula communities altered the structure of the rural economy and the dynamics of rural social institutions. The regime of requisitions discussed in this chapter produced tensions within rural communities that, in turn, influenced the elaboration of subsequent policies. Conflicts erupted between canton chiefs and lineage heads over compensation for forced cultivation, labor and military recruitment, and tax collection. Cleavages also appeared within lineages, as their production units found it increasingly difficult to support lineage members in light of food and labor requisitions. These stresses in rural economic and social relations came to a head in the post-World War II period and will be discussed more fully in chapter 4. Suffice it to say here that these conflicts gestated during the inter-war period and ultimately paved the way for the expansion of cotton in the post-independence period covered in chapters 5 and 6.

The "disguised corvée"

During the first years of colonial rule in northern Côte d'Ivoire (1898–1912), wild vine rubber (*Landolphia heudelotti*) was the principal export crop. With the collapse of the West African rubber trade in 1913, colonial officials were anxious to promote an alternative cash crop. As early as 1898, the Minister of Colonies expressed some optimism about cotton's future in France's new West African colonies. In a letter to the governor general of AOF, he stated that both rubber and cotton "should be the principal sources of wealth" in the region that included northern Côte d'Ivoire.[5] Governor Gabriel Angoulvant had already decided in 1908, during the first year of his eight-year governorship, that cotton would be a viable alternative to rubber. His optimism was in part based upon the reports of colonial botanists and administrators on peasant cotton growing.[6] For example, Maurice Delafosse, the first civilian administrator of the Korhogo region, reported "relatively extensive cotton plantations" in the Guiembé area during a tour of the region in 1907.[7] In some of the earliest photographs of the Korhogo region, cotton is also conspicuous in local markets and communities (Plates 8 and 9).

Governor Angoulvant had been contacted by agents of the Association Cotonnière Coloniale (ACC) (the Colonial Cotton Association), whose mission was to stimulate and assist in the development of cotton in France's colonies in order to reduce the textile industry's dependence on American cotton imports.[8] After his meeting with ACC representatives, Governor

Angoulvant ordered the Agricultural Service in each district to investigate the nature of indigenous cotton cultivation and marketing. He instructed his administrators and agents

to select the best indigenous species, to show natives the superiority of yields in European cotton gins, *to expand cultivation beyond local needs*, all in a way to lower prices which up until now have been too high for export.[9]

New and promising hybrids were selected and exchanged from one colonial experiment station to another. For example, the *hirsutum* variety known as *Allen* was of American origin, and first acclimated in Uganda and then Nigeria. It was introduced into the French Soudan (Mali) in 1925 and Côte d'Ivoire in 1927, where it underwent extensive field trials and was ultimately selected for the new cotton program introduced in the 1960s (see chapter 4). From the start, cotton seeds were selected for their fit with the highly mechanized European spinning industry. Spinning machines were usually calibrated to the American standard medium staple length.[10] Agronomists also considered gin yields to be an important characteristic for selection. The percentage of cotton lint that could be obtained from a kilogram of seed cotton was of great interest to cotton buyers. Summarizing the work of the Agricultural Service during the first five years of cotton research, the head of the agency remarked that it was these two considerations that guided colonial agronomic research.

One must note in conclusion that up until now we have focused our efforts more on selecting a cotton variety that is adapted to a type of (spinning) machine and giving yields that are most advantageous to buyers than on asking ourselves whether the high performing plant was one that brought the biggest economic return to the grower.[11]

Precolonial methods of cotton cultivation involved intercropping perennial cotton with yams, rice, sorghum, or maize in either individual or extended family household fields (*kagon*) and lineage-level fields (*segbo*). Cotton was a secondary crop that did not require much labor time. In yam fields, for example, peasants planted cotton in every fourth mound. Perennial varieties were cut back each year and produced fiber for up to ten years. Cotton yields averaged 40 kilograms per intercropped hectare.[12] At harvest, seed cotton was sold by field owners to Jula traders and elderly Senufo and Jula women. The latter extracted the seeds, carded the cotton fiber, and then spun it into thread.[13] The thread was sold to Jula weavers who bartered their cloth with the Senufo for foodstuffs and engaged in long-distance trade. Locally, a major use of cloth was for shrouds in Senufo funerals. According to their rank, the dead were wrapped in dozens of shrouds before they were buried. Individuals and especially lineage heads accumulated cloth and stowed it away for such occasions. The fact that only locally woven cloth was used in these ceremonies was an important stimulus to the handicraft industry.[14]

A major constraint to increasing cotton production was the long portage required to transport it to coastal ports. The Commandant of the Korhogo region believed that

> when the means of transport are adequate for facilitating the profitable flow of products of the soil such as cotton and maize, the effort required to lead the native to produce for export will be almost null.[15]

When the railroad reached Bouaké in May of 1912, cotton and other crops were swiftly imposed on the peoples of the savanna. Following the instructions of the governor of the colony and the Korhogo district commander, district guards forced peasants to double the size of their rice fields and to plant 500 hectares of cotton.

In 1913 some officials believed that peasants could increase cotton output by intensifying the customary method of intercropping cotton with subsistence crops. The advantages of intercropping were: (1) local varieties resisted parasitic diseases better in association with other plants; (2) the dense plant cover of intercropped mounds maintained soil humidity; (3) food was produced as well as cotton fiber, and (4) the problem of disrupting local farming methods to increase cotton production could be avoided by building upon indigenous cultivation techniques. Thus, in a letter to Governor Angoulvant in September of 1913, the head of the Agricultural Service suggested that peasants be "advised" to reserve two-thirds of each yam mound for cotton plants.[16] This density of cotton plants was significantly greater than the pre-colonial pattern of cotton grown in every fourth mound.

Noted for his disdain of the "native's ways," Angoulvant believed that "scientific" methods of cultivation should be imposed in Côte d'Ivoire.[17] By this he meant monocropping cotton in rows. Monocropping also made it easier for district guards to delimit cotton fields and to supervise their cultivation. Collective fields and payments to village chiefs also facilitated the district commanders' task of verifying that the amount of cotton demanded of each village was in fact marketed.[18] Forced cotton was thus monocropped in fields whose size was determined by the number of taxpayers in each village. The field where compulsory cropping took place became known as "the commander's field." The commander's field was not one large field but a mosaic of fields within the same area, each field cultivated by the members of separate lineages.

In Katiali, a "guard" from administrative headquarters in Korhogo came to stake out the commander's field. He returned periodically to oversee the weeding and harvesting phases of the production cycle. Bêh Tuo, an elderly

Senufo informant, described some of the punishments inflicted on individuals when they failed to perform agricultural work to the satisfaction of district guards. Guards often singled out lineage heads of production units, forced them to lie on the ground, and whipped them while balafon players and drummers played music that was normally heard in hoeing contests. Some were forced to carry heavy rocks on their heads throughout the village.[19] As Isaacman and Likaka demonstrate in their studies of forced cotton in colonial Mozambique and the Congo, such acts of brutality were common where states were weak and unable to supervise effectively the labor process. They were principally aimed at intimidating rural populations to follow orders or else suffer the painful consequences.[20]

Contrary to the view that little effort would be expended to increase cotton output, the forced cultivation of cotton and food crops meant that more labor time and agricultural land was devoted to commodity production during the rainy season – that is, during the period of subsistence production. In the absence of any improvement in farming methods or technology, compulsory cropping meant an immediate reduction in resources available to food production. Given the low level of productivity and the demands on households to furnish laborers for public works projects, one administrator questioned the wisdom of promoting cotton:

> This crop (cotton) is still relatively poorly developed in the district (*cercle*), and it will require a tremendous effort for natives to give it the importance it deserves . . . Subsistence crops take up a considerable amount of time in light of the primitive methods employed and the great number of adults working outside the district, depriving agriculture of as many strong hands; (hence) it will be absolutely necessary to improve (these methods) and means of transport – which will ultimately make available a large part of the workforce and the means of cultivation to increase the cultivated area with the same labor force.[21]

In the minds of colonial administrators, there was a difference between the imposition and supervision of cotton production and forced cultivation *per se*. Rice, maize, millet, yams, and shea nuts were the officially recognized "forced crops," while cotton was considered a "disguised corvée."[22] At the village level, the chiefs of residential quarters were obligated to fill a designated number of sacks full of rice, maize, millet, and peanuts. The number of sacks was determined proportionately to the number of taxpayers in the village. Cotton was not subject to the same method of coercion. Unlike in the neighboring French Soudan (Mali), administrators in Côte d'Ivoire did not specify a number of cotton bales or kilograms that had to be produced.[23] They only fixed the area that had to be cultivated in cotton. In 1916 each taxpayer had to cultivate 0.08 ha; this required area increased to 0.10 ha in 1918.[24]

Farmers were initially given the latitude to choose the site and to clear, mound, and plant cotton fields. However, once seeds were in the ground, a constant surveillance of cultivation methods ensued. District guards and Agricultural Service extension agents visited villages to instruct peasants to thin, top, weed, and, after the harvest, to burn cotton plants.[25] During their tours, district guards inspected fields to see if farmers were complying with orders. If they were displeased with the quality of work, they would force the residents of an entire quarter to return to the commander's field until the work was satisfactorily completed. Those who refused to follow directives were publicly whipped. An elderly informant, Gniofolotien Silué, recalled that such acts of brutality were often directed against lineage heads to set an example. To avoid seeing elders publicly humiliated, lineage members and their dependants worked in the commander's fields and carried the produce to marketing points.[26] Canton and village chiefs controlled any money earned from marketing cotton and other forced crops. When it was sufficient, this money was often used to pay the head taxes of village members.

In 1916, 9,600 ha of cotton were monocropped in the district[27] and 600 men were working in the forest region.[28] The district's population was also forced to participate in the "Food for Victory" campaign being waged in France's colonies to provision its soldiers on the European warfront. This "in-kind prestation" demanded by the Minister of Colonies resulted in the export of 300 tons of maize, 150 tons of millet, 150 tons of paddy, 10 tons of peanuts, and 30 tons of shea nuts from the Korhogo district in 1917. In the context of existing requisitions and the additional burdens of military conscription and the long distance transport of wartime exactions by porters, these demands made severe inroads on local resources.

The intensification of state requisitions in the form of food, cotton, soldiers and labor power did not proceed without some resistance from the local population. In the November 1915 political and economic report of the Korhogo post, the district commander wrote that "certain chiefs" had opposed the commander's orders with the "force of inertia."[29] He noted that the resistance came from the chiefs' belief that the people they were allowed to administer by the colonial government "belong to them." Furthermore, family heads, "with more reason this time, think that we are diminishing their production force by taking from them the instruments of labor."[30] The state's approach to overcoming peasant resistance to forced cultivation was threefold. The first method was to offer attractive prices to producers as an incentive to increase production for the market. The reasoning behind this approach was expressed by Governor General Joost Van Vollenhoven in a letter to the governors of AOF when he wrote:

> The native of AOF is not unlike the rest of humanity. He has come to offer his services and his products each time that one has offered him remunerative prices. In contrast, he refuses to work each time he deems his salary insufficient. One could

say, without exaggeration, that crop yields in AOF were a function, not of climate, but of the prices paid during the preceding harvest.[31]

In contrast to this example of the rational peasant development discourse, cotton producers were rarely remunerated for their labor. The money earned from cotton sales usually went to canton and village chiefs. If higher prices were to work as an incentive to increase production, then individual producers would have to be paid. However, the state's policy of using chiefs as intermediaries, combined with the surveillance advantages of monocropping, perpetuated the practice of collective fields and payments to canton and village chiefs. Thus prevailing production relations required merchants and the state to follow a second, time-tested recourse – the practice of offering incentives to chiefs in the form of credit, bonuses, and commissions, as rewards for intensifying commodity production.

A third approach to increasing yields was to intensify coercion (Plates 10–13). Requiring producers to increase the area under cultivation became a standard practice. But enlarging the area under cotton could not guarantee higher yields. Once fields were sown, they were easily neglected. Hence the state had to heighten its surveillance of peasant cultivation. A report written in 1918 on cotton cultivation in Côte d'Ivoire reveals the degree of state supervision of peasant cultivation in the organization of forced cultivation.

It is an ongoing job for administrators, for not only is it necessary to compel the natives to clear the land and sow, but it is also necessary to hold his hand in the cultivation of fields up until the harvest, and finally to form convoys to selling places.[32]

It was the task of district guards and extension agents of the Agriculture Service "to keep abreast of crop conditions and to monitor its cultivation."[33]

Colonial officials were clearly pleased with the results of their efforts. In a self-congratulating manner, the head of the Agricultural Service, H. Leroide, summed up what he considered to be the sources of increasing cotton exports.

From nothing in 1912, production went from 94 tons in 1915, to 357 tons in 1916, and to 540 tons in 1917 . . . This result is entirely the work of the local administration. It has achieved this by obliging natives to expand their cropping area and by organizing the buying, ginning, and baling of cotton which have led to larger harvests and suitable marketing conditions that have made this textile an export good.[34]

Senior administrators saw themselves as playing an important role in breaking what they saw as a "vicious cycle" in which textile firms were hesitant to invest in ginning and shipping machinery unless they were guaranteed sufficient quantities of cotton at prices low enough to ensure a profit. They believed that "only a powerful and disinterested agent" such as the colonial state could break this cycle, which would lead to both the "creation of new resources in regions lacking exportable products" and the marketing of a raw material badly needed by the metropolitan textile industry. In the preface to Leroide's report, Lieutenant Governor Angoulvant underscored the Agricultural Service chief's point that force was necessary to make people produce cotton in sufficient quantities for commerce.

> In conclusion, I want to make clear that in a new country inhabited by primitive and not very hard-working populations, one cannot expect to have quick results; success can only be the fruit of long and patient efforts. It is a war of attrition that we must lead against the ignorance and inertia of men, against the bad forces of nature; victory will belong finally to those who have endured.[35]

The major elements of the compulsory development discourse are readily apparent in these colonial texts. It begins with data on the upward trend in cotton exports which are explained as the outcome of the interventions of outsiders. Growers are portrayed as poor (without resources), ignorant, and lacking sufficient technical skills to improve production. They are shown to need external technical expertise which state agencies are prepared to deliver – forcibly if need be. The Agricultural Service views itself as a neutral agent of the state whose goal is to create the conditions of cotton development for export. By breaking the so-called "vicious cycle" of low cotton output, the Agricultural Service is thus at the service of development. But development for whom? What distinguishes this early expression of the development discourse from later variants is its transparent colonial objectives and methods. Clearly the main beneficiaries of cotton development are not the peasant farmers who are being coerced to grow the crop. It is the French textile industry and the metropolitan state that are to benefit. The former hopes to obtain a cheap raw material while the latter seeks to free itself of costly and fluctuating American cotton imports. As the Agricultural Service chief made clear, the primary goal of cotton development is to "assist effectively the Mother Country" by furnishing the metropolitan textile industry with its basic primary material.[36]

Renewed coercion was the approach used in 1918 when the imposed area under cotton was increased to 0.1 ha per taxpayer. This measure was used in the wake of the steep drop in cotton sold to commerce in 1917 (Fig. 3.2). District administrators attributed the sharp decline to a number of factors: first, monocropping made plants more vulnerable to pest invasions; second, the metropole's request for unlimited food supplies at more remunerative prices than cotton resulted in a neglect of cotton fields; third, forced labor recruitment for public works projects resulted in some cotton fields being abandoned; and fourth, due to the higher costs of imported cloth during the war, local weavers had increased their production. Most importantly, the Jula merchants who supplied these weavers offered higher prices to producers for their cotton.[37]

The parallel cotton market

Competition between indigenous traders and European merchant houses for local cotton was a continual concern of administrators and representatives of the metropolitan textile industry. Colonial administrators recognized the importance of local cotton markets in 1913, when the official price for seed cotton was aligned with regional prices.[38] To capture the cotton that supplied Jula and Baoulé weavers, a concerted but ineffectual effort was made by the state to suppress the indigenous textile industry. One method tried was to limit the use of manual cotton gins by Jula traders (Plate 11). In 1915, the head of the Agricultural Service unequivocally stated the official position when he wrote:

> If we are seeking to develop cotton, it is with a view towards provisioning an export trade with a primary material while facilitating the import of European cloth. Hence, the native must be disposed from the outset to deliver his cotton to commercial houses so that a gradual suppression of local weaving will result. The aim of the portable (ginning) machines is precisely to permit the processing of seed cotton on the spot into a form that makes its transport economical. Thus, the natives must not become accustomed to using them only for processing cotton that they sell to weavers.[39]

How and whether the Agricultural Service implemented this policy and the responses of local cotton growers and traders are not apparent in the archival record. What is clear is that the demand for cotton by local traders significantly increased during the First World War years. A resurgence in local weaving occurred in 1917, following a tripling in the price of imported cloth. European cloth that used to sell for two francs was then selling for six francs. The demand for locally made cloth was so great that weavers paid one franc per kilogram for cotton fiber in villages, while merchant houses offered 0F.35 per kilogram at great distances from the point of production. This combination of higher prices and shorter distances made local markets more attractive to cotton growers.[40]

In a special report examining the reasons behind the steep drop in cotton sales to commercial houses between 1916 and 1917, Governor Raphaël Antonetti remarked that the revival in the handicraft industry seemed logical under the circumstances.

> Depriving producers of their harvests was especially disastrous; it was tantamount to killing a particularly prosperous local industry that, to the contrary, should have been encouraged: the making of cloth. The bit of cloth of European origin which before cost 2 francs, now sells at 6 francs. Indigenous cloths are thus highly sought after; so much so that weavers do not hesitate to pay 1 franc per kilogram, hand-ginned, right in villages, the product that the mill, often very far, only pays 0F.116 for seed cotton, or 0F.35 per kilo of ginned cotton. One can easily understand in these conditions why the native is no longer interested in the cultivation of cotton.[41]

In response to Governor Antonetti's report, Governor General Angoulvant reiterated his wish to suppress the indigenous handicraft industry.

I do not share your opinion of encouraging cloth making by native weavers, indeed, just the opposite. These petty craftsmen spend a considerable amount of time making an insignificant amount of fabric and the women spend all of their days spinning. They both could make much better use of their time cultivating by expanding, for example, their cotton fields. Everyone would benefit then, the natives as well as European commerce.[42]

Angoulvant held no illusion as to what it took to increase cotton production when he wrote to Antonetti, "if you had been informed that . . . the results have been and will remain subject to administrative pressure, you could have obtained a few more tons."[43] Antonetti knew quite well what it took to increase cotton output. His priorities, however, lay elsewhere. His major concern was maintaining a steady flow of labor from the savanna region to European timber concessions and plantations in the forest region. According to the president of the ACC, Antonetti did not want to exert too much pressure in the cotton growing areas for fear that it would lead to a reduction in labor needed in the forest area.[44]

Such disputes among senior administrators reveal fundamental differences in how colonial authorities perceived their mission and the place of Africans within it. They also illuminate the dilemmas faced by colonial officials as they tried to reconcile the conflicting interests of the metropolitan textile industry and the expatriate population with those of the African population. Côte d'Ivoire Governor Antonetti clearly sought to balance the labor needs of expatriate investors located in the forest region with the labor needs of cotton-producing households in the savanna region. He also faced the dilemma of whether to suppress or encourage local handicraft industries, and the effects such policies would have on cotton production. Confronted with such policy questions, it is not surprising that colonial administrators adopted contradictory policies. African peasants and merchants occasionally found some room to maneuver within this contested policy-making environment.

The latitude producers had in selling their crops was crucial to the existence of the parallel cotton market. Peasants could sell cotton to whomever they wanted. The state only engaged in commerce when there was "insufficient free trade."[45] Its role was to put the indigenous producer in contact with merchants "so that he does not experience any insurmountable difficulties in disposing of his cotton and that he not be forced to transport his products over too great a distance in order to sell them."[46] The problem of distance from point of production to market was still important in 1922, when cotton from some villages in the Korhogo region had to be head loaded 200 kilometers by porters.[47] Moreover, imported cloth prices

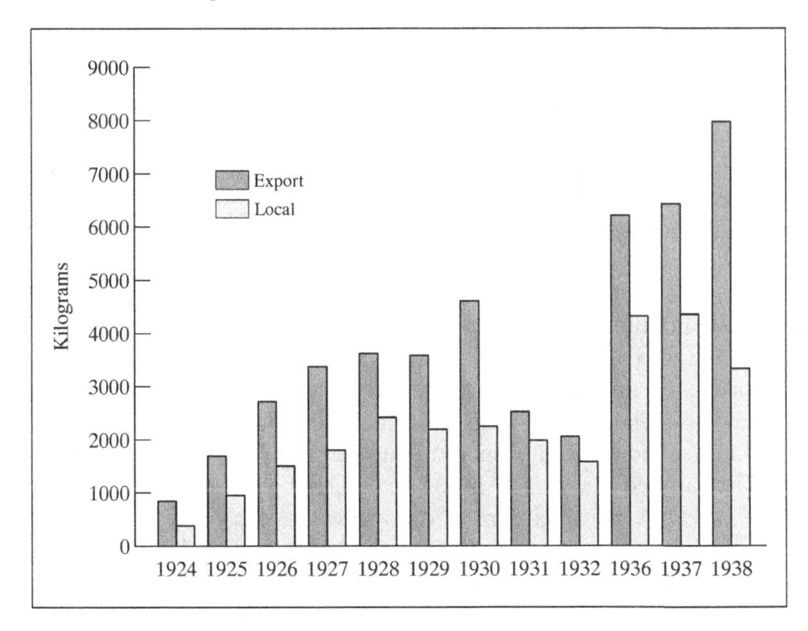

Fig. 3.3 Seed cotton sales in local and export markets: 1924–32; 1936–38 (*Source:* ANRCI 1 RR 67c)

continued to be extraordinarily high as France's economy sank into a deep recession. Cloth selling for nine francs in 1918 sold for eighty francs in 1920. Cotton export prices were depressed, and the local weaving industry boomed.[48]

The situation remained unchanged in the early 1920s. During a tour of the cotton-growing areas of French West Africa in 1924, Arthur Waddington, the president of the ACC, remarked that the indigenous textile handicraft industry consumed a large part of the cotton produced in the colonies. To increase cotton production for export, he suggested that the colonial state impose a tax to be paid in seed cotton. The administration, he reasoned, would not have any problem marketing its cotton and would be assured of a reasonable purchase price.[49] Waddington's cotton tax was never imposed. Governor Antonetti viewed the parallel cotton market as essentially a price problem. In 1925 the export market rebounded when merchant houses offered 1F.75 per kilogram of seed cotton – ten times the price it offered during the war period. Nevertheless, producers were still selling more than one third of their cotton to local traders who offered 3F.60 per kilogram.[50] For cotton sales to increase for the export market, this price gap had to be narrowed.[51]

Figure 3.3 shows that throughout much of the 1920s and 30s, the parallel cotton market thrived in Côte d'Ivoire as it did elsewhere in French West Africa.[52] Although the ACC data for cotton exports were reliable, administrators could only guess at the amount of cotton sold in local markets. On the basis of population densities and assumed consumption levels the figures for local sales represent best guesses.[53] Working with these numbers, an estimated 37% of total cotton output was marketed locally. These percentages were higher when the price of European cloth rose and demand for indigenous cloth spurred local market sales.

The colonial state took a step towards closing the price gap in 1928 when it established a minimum purchase price for cotton below which producers were advised not to sell their cotton. Some administrators viewed European cotton traders as unscrupulous because their purchase prices were well below the world market. As a result of these low prices (e.g. 0F.60/kg in 1927), officials had difficulty encouraging peasants to grow cotton for export. They argued that prices should be determined by market rates prevailing at Le Havre minus ginning and transportation costs. Over the protests of the chamber of commerce and ACC, who called for "free commerce," the cotton price was set at 1F.75 for 1928.[54]

The push for cotton exports

While the debate over purchase prices continued, administrative coercion was stepped up to increase cotton output. Beginning in 1925 a new set of policies was introduced that deeply affected production and exchange relations in the cotton growing areas. On the instigation of Governor-General Jules Carde in Dakar and Governor Maurice Lapalud in Côte d'Ivoire, and under the supervision of the newly created Textile Service, production quotas were fixed for each district. As a result, villagers were required to double the size of their cotton fields.[55] Bonuses were also offered to chiefs to raise cotton quality and quantity. Head taxes were raised to 7F.50 – more than double the levy of 1913. Governor Lapalud also intensified the recruitment of northern men to work in the forest region.[56]

Further measures were taken in the Korhogo district itself by Commandant Lalande, who in 1927 directly intervened in the marketing of peasant production. In concert with the district's merchant houses, the administration established regional market "fairs" (*foires*) where peasants were required to sell their produce. They were organized throughout the colony and much of West Africa on the instigation of Governor-General Carde, ostensibly to shorten the distance peasants had to carry their cotton to market and to ensure that growers received a fair price for their crop.[57] Carde also viewed the fairs as a means to impose quality standards by penalizing growers who failed to sort their cotton into high- and low-quality

Plate 8 Local cotton marketing at the Niellé weekly market (*c.* 1910)

Plate 9 Spinning cotton thread (*c.* 1902)

Plate 10 Cotton market at Tiébissou (*c.* 1915)

Plate 11 Operating a manual cotton gin in Bouaké (*c.* 1915)

grades. Although the implementation of the fairs varied within as well as between colonies, they all shared the common element of coercion in the guise of free markets. The contradiction did not go unnoticed by the highly regarded French journalist Albert Londres, who traveled in francophone Africa for four months in 1928. One of his stops was in Bouaké, in central Côte d'Ivoire, where he reported with considerable irony the following scene.

Today, the cotton carriers are hurrying toward Bouaké. They go most willingly. Although burdened by their loads, they still smile at you. The guard acts like a sheep dog. Men, women, children, everyone is driven to *corvée*. They will not see their homes for four days but they are going to see the commandant! For those who bring poor quality cotton, they will get, as a matter of course, a few days in jail. They are speeding up!

They reach town. They file into the courtyard of the commandant, put down their sacks and open them up. They stand in line. The guard admires his work, and then walks in an official manner to the office of the god of the bush. He freezes at the doorway, and with his hand at his fez, thunderously announces:

"Commandant! the cotton is here!"

Plate 12 Forced cotton deliveries, Bouaké (*c.* 1928)

Plate 13 Forced cotton carriers on the road to the Bouaké cotton market (1928)

Plate 14 Forced labor on the Côte d'Ivoire railroad

Plate 15 Peanut sacks on the wharf at Grand Bassam

Plate 16 Forced laborers loading logs onto railroad car near Azaguié, Côte d'Ivoire

Then the ceremony began. The commandant left his office and began to examine each sack.

"*Atakouré*!" he says. "*Atakouré*! Very good! very good! Go weigh it."
The lucky ones head toward the *official* scale.

"Go sort it!" he says to someone in an incensed tone.
A man steps out of line, empties his sack and separates the good from the bad.

"To jail!" the commandant suddenly cries. "To jail with you!" he says to the next person.

The cotton of these two is not of good quality; they walk to jail by themselves. Do they understand that it is their fate?

At the scale, they do the accounts. Today, cotton sells for two francs-forty. The clerk marks on a small piece of paper: 26 kilos at 2 fr 40 = 62 fr 40. It is the sum that the buyer must pay them. Confidence reigned!

The guard stood straight in the middle of the yard, seeming to wait for something.

"Sanna," the commandant says to him, "you have done good work; you shall have fifty francs as a reward."

The fifty francs must please him, but it is especially the honor that subtlely transports him. He feels uplifted and his heart is pounding!

"Thank you, commandant!" he shouts. "Thank you! Thank you! Thank you!"

Now the Negroes leave for town to sell their goods. In the business district, the merchants are in front of their counters. They have an ironic look. They seem to say: "Do you see how the administration treats us whites?" If a black clerk goes running after a cotton seller, off to jail he goes. But they are allowed to call out to the sellers.

"Hey, hey there! Come over here. Sixty-two-francs-forty?" yells a white man. "I will give you sixty-four francs. Sell me your sack."

The Negro flees. He wants sixty-two-forty, not sixty-four.

Is there anyone on earth with better instincts?[58]

Merchants commonly underweighed produce and short-changed peasants. Farmers had no choice; they had to sell their products. If they had insufficient cotton to sell, they were forced to buy it from other growers, often at high prices. When they turned around to sell this cotton at "fairs", cotton-deficient households commonly received just a third of the price they had just paid.[59] District commanders condoned such practices. In a letter to the head of the Textile Service, the commander of the Baoulé district based in Bouaké stated:

I decided that unless instructions were received that were contrary to your policy, I would close my eyes to these dealings because I assured myself that if certain natives are obligated to buy cotton from among their own people, it is not due to unfortunate circumstances, independent of their will such as floods or pest invasions, but due to their laziness encouraged by the hope that the administration will overlook it. I consider that the penalty that they thus inflict upon themselves will spontaneously constitute the best sanction and that next year it will bear its fruits.[60]

The squeeze was effective. Total seed cotton production in the colony increased by nearly fivefold between 1924 (1,256 tons) and 1928 (6,005 tons) (Plates 12 and 13).

The conditions of peasant production and exchange continued to worsen towards the end of the decade. In April of 1928, the Textile Service instructed district commanders to force peasants not only to monocrop cotton in the "commander's field," but also to intercrop cotton in all food crop fields "without exception."[61] Rural labor supplies were strained even further with the arrival of the railroad at Ferkéssédugu in 1928. A number of Europeans, dissuaded from investing in cocoa at the current low market prices, bypassed the forest region to establish sisal, kapok, castor oil, and sesame plantations in the Ferké and Korhogo regions. These northern planters, who by 1932 had been granted 12,000 ha in concessions, received subsidies from the state in the form of production bonuses and, most importantly, cheap labor recruited from within the Korhogo district.[62] In 1930 over 1,000 forced laborers were working on these plantations.[63] Furthermore, in 1929, Governor Lapalud encouraged commanders of the northern districts to force peanut production on the local population in order to supply a recently constructed peanut oil factory in the Ferkéssédugu region. Lapalud encouraged the commanders "to push" peasants to produce 4,000 tons of peanuts for the factory.[64] Head taxes were raised again in 1930 to 12 francs. In the same year, official cotton prices fell and then dropped dramatically to 0F.65 per kg in 1931.

The extraordinary drain on household resources in the Korhogo region by 1930 starkly contrasted with the relatively favorable conditions that prevailed during the early 1920s. In that ten-year period, there had been a thirteen-fold increase in forced labor recruitment, head taxes had doubled from 6 to 12 twelve francs, and the area under cotton had doubled; by 1930 peanuts were forcibly cultivated in the Boundiali and Korhogo regions, cotton and peanut prices were low, and the markets were monopolized by European merchants who set their own standards.

This unprecedented regime of requisitions put household subsistence security at considerable risk. A severe drought hit the north in August of 1930. A locust invasion accompanied the drought, further lowering the yields of the long-cycle crops – rice, millet, and cotton. Combined with the forced cultivation of cotton and peanuts and low market prices, the drought and locust invasions triggered an already explosive situation. Severe food shortages struck the entire district, hitting the Boundiali and Korhogo subdivisions the hardest. A plate of rice sold for 2 francs in mid-1931 in Kouto, north of Boundiali, and there were reports of people indenturing themselves as domestic slaves for 100 francs.[65] The canton chief of Boundiali further aggravated the situation by guaranteeing the merchant house de Tessière a large part of the canton's 1930 rice and millet harvests before seeds were even planted.

A Ministry of the Colonies Inspection Report of 1931 strongly criticized the Korhogo district's administrators for their "extravagant interventionism and ignorance of local economic possibilities." They were also reprimanded for their "suppression of free commerce" and "forcing the indigenous population to cultivate cotton and peanuts without ever asking itself if these crops were really adapted to the Kong [Korhogo] District." Inspector Haranger wrote:

If we wanted to go into details on all of the irregularities and abuses committed in the district of Kong, a simple report is insufficient; a volume would be necessary . . . It will suffice for the moment to point out that all of the food reserve granaries have been emptied, and in some cases broken into, and all that remains for the people is their eyes with which to cry . . . The final result obtained by this method, worthy of the pitiless hordes of the former black conquerors, is famine everywhere in this country.[66]

Inspector Haranger provided plenty of evidence to support his accusations of "administrative terror" in the region.[67] He conducted investigations in more than fifteen communities in the Korhogo and Boundiali subdivisions. One of his stops was in Katiali.

Katiali (canton center). – The inhabitants suffered from hunger in 1930. Locusts ravaged the crops. The village had to hand over all that remained. It delivered to the fairs a truck full of rice that was taken from food reserve granaries by district guards using force. The Canton Chief (Ténéna Silué] and a person named Béma Koné were beaten with fists by the guard Zié Koné. Béma Koné was even tied to a tree while they emptied his granary. The canton had to buy rice to feed itself.

It furnished to M. de Chanaud [owner of the peanut oil factory] 41 sacks of peanuts. It received, long afterwards, 395 francs. It handed over to requisitions 30 chickens and 80 guinea fowl to Malimadou Coulibaly, son of Chief G'Bon Coulibaly and head of the [*jasa*] representatives. Neither the chickens nor guinea fowl were paid for.[68]

Open resistance to the regime of requisitions occurred in the canton of North-Niéné in the Boundiali subdivision, where villages had surrendered most of the foodstuffs and "had not been paid, but had been beaten a lot."[69] The "Tengréla revolt," as it became popularly known, took place over the hungry season of July–August 1930. A detachment of district guards and irregulars (*goumiers*) were sent on 21 August 1930 to suppress this "display of ill will" towards the colonial state. In the midst of the political and social crisis, Commander Lalande was dismissed from his post and the canton chief of Boundiali was publicly reprimanded; both were severely criticized for "not taking an interest in the people they administrated."[70]

Following the food crisis of 1930, there was no cotton campaign, "properly speaking," in the Kong district in 1931 nor in 1932.[71] Nevertheless, 125 metric tons of cotton were sold on the export market in 1931 and 75 tons were consumed locally. The tonnage for 1932 was slightly higher: 162 tons for export and 75 tons for local markets.[72] Production totals amounted to one-quarter of the tonnage marketed in 1927 and 1928. Official market prices were extremely low, with a kilogram of cotton selling for 0F.65. As Governor Reste exclaimed in a letter to Governor General Jules Brévié in August of 1932, "Present selling prices are veritable famine prices."[73]

The tonnage sold in local markets is most likely underestimated. The Textile Service was able to calculate the tonnage sold on the export market from records kept by cotton gin operators in various districts. However, since most of the cotton marketed locally was ginned by hand in villages, it is difficult to know exactly how much cotton was sold in indigenous markets. In a report by des Etages, an engineer responsible for cotton trials in the Kong district in the early 1930s, it was noted that two-thirds of the cotton grown by the son of the canton chief of Korhogo was sold locally for a much higher price than the official rate.[74] As director of the "Native Model Farm" of Korhogo in which cotton was grown, des Etages closely followed the production and marketing of the farm's products. Yet, outside of this farm, administrators did not know how much cotton was produced and marketed locally. They simply guessed that 1 to 1.5 kg of cotton "per inhabitant" were sold locally.[75]

At the cotton experiment farms in Ferkéssédugu and Bouaké, field trials were underway with a new cotton variety known as Ishan (*Gossypium vitifolium*) imported from Nigeria in 1928. In 1933 cotton monitors employed by the Agricultural Service began to replace indigenous varieties with Ishan in the Bouaké, Dabakala, and Séguela districts. While the district commander of Korhogo, Michel Perron, waited in anticipation for the improved variety, he stated that "for the moment, we must content ourselves with preventing the native from losing complete interest in cotton."[76]

Cotton production substantially increased under administrative coercion in the Korhogo district between 1933 and 1935. The tonnage of raw fiber produced in 1935 equaled the quantity produced in 1929. The compulsory extension and cultivation of the higher-yielding Ishan variety accounted for much of the increase in production. Food shortages were again reported in the Boundiali region in 1933 by M. Bordarier, the commander of the Korhogo district. Bordarier expressed little sympathy for the subsistence problems plaguing the local population when he wrote:

> I want to digress here on the particular apathy of the natives of the Boundiali Subdivision who "defend themselves less" before the demands which are made on them (than) the natives of other Subdivisions who carefully put aside plenty of foodstuffs necessary for the hunger season ("la soudure").[77]

In 1936, after an ambitious campaign to extend the new varieties throughout the cotton-growing regions, great losses were reported from insect pests and parasites. Over 16 tons of improved Budi seeds[78] had been distributed in the Korhogo, Boundiali and Ferkéssédugu regions that year, and only 20 tons of cotton were harvested – 13% of anticipated yields.[79]

Following an inspection of the Ishan cotton-growing regions south of the Korhogo district, des Etages recommended that cotton be intercropped with food crops as a means of pest control. He noted that the bulk of cotton produced that year had come from intercropped fields. No less important, des Etages recommended intercropped fields because "they demand a lot less labor from the natives while furnishing them foodstuffs and products to sell."[80]

The savings in labor-time devoted to cotton cultivation was one of the advantages mentioned by Marcel de Coppet, the Popular Front governor general of French West Africa, when he recommended a return to the perennial cotton variety: "It is not doubtful that the cotton shrub can be cultivated over the widest area because the cultivator finds himself largely freed, for a few years, from the considerable labor of preparing the soil for sowing."[81]

Partisans of the rational peasant development discourse, the de Coppet government argued that cotton production would only increase when first, it became equally, if not more, profitable as rice and millet, which gave a higher return to labor; and second, when the considerable differences between local and export market prices were eliminated. Otherwise, peasants would market more remunerative crops and sell cotton in local markets. "Thus before being able to export (cotton)," one official pointed out, "the local market must be saturated."[82] Frederick Cooper suggests that the pro-peasant position of the Popular Front government in 1936–39 was not uniformly shared by lieutenant governors and district commanders who continued to condone the recruitment of forced labor to French settler plantations. Such disagreements among senior colonial officials over the labor question appear to have provided peasant producers with some latitude in the marketing of their crops, at least for a short time. The resurgence in forced cultivation under the Vichy government during the early years of World War Two rapidly eroded this modicum of peasant latitude in rural markets.[83]

The decline of cotton

The period 1938–46 encompasses both the zenith and nadir of cotton exports from colonial Côte d'Ivoire (Fig. 3.2). The Ishan variety had by now been introduced throughout much of the central and northern savanna regions. To maintain its relatively high yields and quality, greater control was exerted at ginning stations and in farmers' fields to prevent its hybridization with local varieties. The results were impressive. In 1938, 6,640 tons of seed cotton were harvested in the Bouaké, Dabakala, and Séguela districts alone. An estimated 11, 219 tons were harvested in the entire colony. Exports of cotton fiber reached an all-time high (2,535 tons). The results were "extremely encouraging;" a breakthrough seemed to have taken place.[84] But this did not turn out to be the case. The measures taken to control the quality of Ishan seeds were relaxed when the longtime director of the Textile Service, M. Jacquier, went into retirement in 1938. Production quickly plummeted in the following years, reaching an all-time low in 1946. More than a few administrators were disillusioned with the trend. J. Lebeuf, the former director of the Ferkéssédugu cotton farm and current head of the Agricultural Service, declared in 1941 that Ivorian cotton policies had been a dismal failure.

It seems in effect that a major error was committed around the years 1920–25 that still weighs heavily today on the agricultural economic orientation of the soudanian and voltaic regions. Just because cotton was found to grow everywhere, one concluded that it could become an important agricultural product of these regions. A hypothesis even more seductive since this textile is one of the primary materials for which the Metropole depended almost exclusively on foreign sources and for which it paid a heavy tribute. The reality appears, alas, something entirely different.[85]

Over the opposition of some administrators and the Agricultural Service, a cotton conference was held in Bouaké in 1940 to address the downward slide in cotton.[86] The meeting was organized by Vichyite Governor Horace Crocicchia at the request of local textile interests and the Chamber of Commerce. The Governor expressed his commitment to the cotton sector and ordered a renewed effort in its favor. He also sent the reluctant director of the Agricultural Service to Nigeria to obtain Ishan cotton seeds, to replace the degenerated stock for the 1941 cotton season.

Governor Crocicchia's initiatives were complemented by the wartime reorganization of the cotton sector within all of French West Africa. In 1941, Pierre Boisson, high commissioner of Vichy-controlled West Africa, promulgated a series of laws aimed at improving the quality and quantity of cotton in West Africa.[87] His Convention Cotonnière Coloniale (Colonial Cotton Plan) gave local administrations new regulatory powers in the cotton sector. The Convention's two basic goals were to offer cotton growers an attractive market price, and to create an extension service that would intervene in the areas of cotton production, ginning, and marketing. To make cotton a more remunerative crop, Boisson recommended fixing prices at a level that would be competitive with other cash crops, such as peanuts. The recently formed Le Comité d'Organisation de l'Industrie Textile (Planning Committee of the Textile Industry) agreed to buy all cotton sold in export markets.

The newly created L'Union Cotonnière de l'Empire Français (UCEF) (Cotton Union of the French Empire) was the successor to the ACC. It was given the role of managing cotton experiment stations, selecting and distributing seeds, and ginning and classifying cotton. Its agents were to serve as technical consultants to district administrators and to cotton monitors trained by the Agricultural Service. The Convention decreed that all cotton destined for export markets be sold and classified in special market centers at which a director would enforce cotton quality standards. It allowed for the sale of cotton "between natives" for local handicraft industries. All cotton sold and ginned at official market centers had to be exported. Jula traders, therefore, were denied access to UCEF-managed gins if they planned to sell fiber in local markets. The Provident Societies of each district were to hire cotton monitors to oversee virtually every stage of the production process: selecting and measuring sites for cultivation; overseeing the clearing, hoeing, weeding and thinning of fields; and finally, being present at harvest time to see that cotton was harvested in a timely manner.

Elders in Katiali recounted how a district guard, rather than a crop monitor, visited the village to oversee these tasks. He came to stake out fields and to oversee their preparation, and he returned later for the weeding and harvest periods. Each residential quarter of the village had its own field to cultivate. Everyone worked in them: men, women, elders, and children. Men usually cleared land and hoed fields for planting; women and children most often planted and weeded fields. Everyone participated in the harvest. Mondays and Fridays, the customary rest days, were designated work days in the commander's fields. If the guard decided that these fields needed attention, people were forced to work in them regardless of the day of the week. Gniofolotien Silué recalled: "If there were too many weeds, the district guard would force us to weed the field. If someone refused to weed, he was whipped by the guard with a branch. I was beaten once . . . The guard's name was Tiéba."[88]

The renewed administrative effort to expand cotton produced a momentary upsurge in exports, with a peak occurring in 1942 (Fig. 3.2). The following year, severe food shortages were reported in the Boundiali subdivision of the Korhogo district. A 1943 Administrative Affairs report by Inspector de Gentile offers a glimpse into the hardships experienced by families in the region in the context of the wartime regime of requisitions. Between 1938 and 1943, individual taxes had quadrupled. In 1943 special wartime taxes were also levied which amounted to an additional 100 F per taxpayer. More than 1,800 men were drafted into military service, and every six months 6,000 young men were forcibly recruited to work on European plantations in the forest region.[89] With rice imports from Indochina halted, the population of the Korhogo district was ordered to furnish rice (3,200 tons), millet (790 tons), maize (505 tons), and 60 tons of wild rubber. In all, 800 hectares of cotton were also imposed on the district. According to

Inspector de Gentile, not enough rice was grown in the Boundiali subdivision to meet the rice corvée quota. Some families were forced to exchange part of their millet reserves to obtain rice from villages in the French Soudan.[90] Famine was reported in every canton of the subdivision in 1943.[91] The colonial state eventually distributed more than 14 tons of millet and maize, 37 tons of rice, and 40 tons of peanuts in emergency food aid.[92]

District commanders reported that it was becoming more and more difficult to rule an increasingly beleaguered population.[93] If the Korhogo district commander's visit to Katiali in May 1942 is at all indicative, the level of discontent was very high. Upon arriving in Katiali he discovered the commander's special quarters to be in disrepair. He immediately called a village-wide meeting, but had to wait two hours before the Jula chief and notables arrived. The commander learned from his interpreter that Katiali had been in a state of "quasi-insubordination" since the late 1920s, following its attachment to the Canton of M'Bengué (its pre-colonial enemy).[94] Since that time the chief of Katiali had not heeded orders coming from the canton chief. The commander responded to all of this by ordering the entire population of Katiali to repair his camp. He next ordered thirty young men to work on the road linking Katiali with Korhogo, which was in "urgent" need of improvement. As a final punishment, he ordered the village chief, the Jula *almani*, and two quarter chiefs to return with him to Korhogo for "administrative retraining."[95]

After 1942, cotton production and exports declined precipitously. From 6,134 tons in 1941/1942, cotton output fell to a mere 56 tons in 1946/47. Imports and exports approached zero as the war in Europe effectively cut off German-occupied France from its African territories. The Chamber of Commerce representative in Bouaké listed additional reasons behind the fall in production: declining cotton prices and buoyant prices for staple foodstuffs, the drain on agricultural labor associated with heightened labor recruitment and grain requisitions, and pest problems.[96] It was estimated that a hectare of yams was 40 to 50 times more profitable than a hectare of cotton in the Bouaké region. In light of this competition between food crops and cotton:

> it is very improbable – even with a maximum increase in yields – that cotton can become some day a crop that will be grown without administrative coercion, which many Administrators find repugnant, the greater the difference (in prices between food crops and cotton).[97]

The "repugnance" felt by district administrators in imposing cotton was also a major factor in the downturn. Partisans of the Free French movement, led by the exiled Charles de Gaulle, viewed forced cultivation as inconsistent with their more democratic and assimilationist vision for

France's African colonies.[98] Both the head of the Agricultural Service and the Free French governor of Côte d'Ivoire, André Latrille, were opposed to forced cotton cultivation.[99] In the absence of administrative coercion, cotton output declined. Appeals made by the Chamber of Commerce in the mid-1940s to resurrect the defunct Textile Service went unanswered by a succession of governors.[100]

The wartime revival of the indigenous weaving industry was a final factor in the precipitous drop in cotton exports. Constraints on maritime shipping led to a scarcity of imported cloth in the colony. Consequently, local cloth became highly desirable and Jula traders paid from double to triple the amount offered at export markets for a kilogram of cotton.[101] In the 1944 annual report of the Agricultural Service, local market prices were four times the export market price. Less than half of the cotton produced in the Bouaké district was sold in official markets.[102] The power of district guards had considerably diminished. The best they could do was provide baskets to cotton growers whom they hoped would deliver their product to export markets. By the 1946/47 season, forced labor had been officially abolished. In the 1946 annual report of the Agricultural Service, the description of cotton marketing in the Korhogo region showed the indigenous handicraft industry dominating the cotton sector.

> The quantity of cotton passing through markets has been very small. The Korhogo mill processed 13 tons of cotton fiber. The area under cultivation remains important, the natives marketing among themselves an important part of the harvest which is spun by women and out of which local artisans weave bands that are used to make clothes.[103]

Conclusion

The image of cotton cultivated everywhere yet rarely sold in export markets is a fitting closure to this period of cotton history in colonial Côte d'Ivoire. Not only does it illustrate a thriving handicraft industry and the persistence of a local cotton market that had supplied that industry since pre-colonial times, it also confirms Roberts's important point that "[t]he French could exhort, cajole, coerce, and punish, but they could not control the decisions and intentions of Africans, who as peasants or as laborers determined the success of colonial policy."[104] The failure of the colonial state to capture a large part of local cotton production for metropolitan industries was largely due to contradictions in colonial policies which created openings for the indigenous handicraft industry and the local cotton market to persist. Only when compelled through administrative coercion would cotton growers produce for the export market. When coercion let up, cotton exports fell dramatically.

One key factor in this scenario was the price difference between local and export markets. It was clear to most observers that the price gap had to be narrowed before output would increase. Yet export prices were partly subject to supply and demand on the world cotton market, over which colonial administrators had no control. Prices were also influenced by expatriate and African merchants, who greatly profited from low purchase prices. Significant price differences between staple food crops and cotton further reduced peasant interest in cotton cultivation. Tensions between the village chiefs and the heads of lineages and extended-family households over the distribution of cotton earnings also influenced the level of cotton output and agricultural policies. Before the advent of the commander's field, cotton was grown in the fields of extended family households (*kagon*) and lineage heads (*segbo*), who profited from its sale in local markets. Payments from forced cotton went directly to the village or canton chief after passing through the hands of the canton's representative in Korhogo. *Jasa* representatives were reputed for skimming off some of the meager earnings from forced cultivation for their personal benefit. Administrators recognized that there was little economic incentive for individuals to engage in cash cropping if they were not compensated for their efforts. However, the logistical difficulty of surveilling cotton growing in geographically dispersed fields made some administrators favor collective fields of monocropped cotton. Advocates of the rational peasant development discourse promoted cotton in extended households' fields and individual payments. These tensions among and between cotton growers and colonial officials combined with uncompetitive prices and a host of agronomic problems (uneven rainfall, poor soils, pests, and low-yielding cotton varieties), help to explain the vagaries of colonial policies and production on the ground at the end of the Second World War.

The reluctance of cotton growers to sell their product in export markets ultimately made officials recognize the economic rationality of peasant production. The attempt to fix cotton prices at levels that were competitive with peanuts in the early 1940s is an example of how peasants' relative autonomy in cotton marketing affected agricultural policies. With the end of forced labor in 1946, the state and metropolitan textile interests had to contrive a new set of cotton policies which included significant price increases. The agrarian history of cotton over the period 1912–46 shows that the parallel cotton market played an important part in the shaping of these policies.

4 Repackaging cotton, 1947–1963

The cultivation of cotton remains a political and price problem.
Agricultural Service, *1948 Annual Report*

The end of forced labor and the total collapse of the export cotton market compelled textile interests and administrators to rethink colonial cotton policy. It was evident to all that the single most important reform had to be improved purchase prices. Unless producer prices were raised to more attractive levels, peasants were not expected to increase their output. Linked to the market price question were the related problems of cotton quality and yields. To receive higher prices, cotton growers had to produce a high-quality product that merchants could sell to French brokers and textile firms. If yields could be increased, incomes would correspondingly rise, even if purchase prices stagnated. The postwar period is marked by the efforts of the metropolitan and colonial states, with the assistance of French cotton research, extension, and trading companies, to erect a new institutional and organizational structure, and to introduce a new set of incentives to overcome peasant resistance to cotton. The new policy was articulated within an evolving development discourse that argued that the expansion of cotton would improve the standard of living of cotton growers and reduce income disparities between the savanna and forest populations. Echoing earlier discourses, cotton development was largely seen as a technical problem that could be overcome by fixing a range of quality, production, and market problems. Peasants were now widely viewed as rational producers who would willingly increase cotton production once these problems were satisfactorily solved. However, colonial officials underestimated peasant resistance to cotton. Its cultivation as a sole crop or monocrop was synonymous with forced cultivation. Not surprisingly, peasants opted for other sources of income, such as commercial food cropping and employment in the coffee and cocoa plantations of the forest region.

The expansion of commercial grain production and labor migration was conditioned on far-reaching changes in the socio-cultural relations of production. Notably, the development of new economic needs and inter-generational conflicts over the control of cash incomes encouraged many young men and some women to emigrate to urban areas as well as to other rural areas. The breakup of large units of production and the cultivation of personal fields by individuals belonging to smaller production units coincided with increased commercial production. On the eve of the cotton revolution, this reorganization of production units along more individual lines became an important condition for the intensification of cotton.

The discipline of the market

If peasants were to increase cotton output for the market, then export market prices had to be as competitive as the parallel market price. However, to get the best price on export markets, cotton quality had to be at least as good as the standard "strict middling" in terms of fiber length, width, strength, and homogeneity. During the era of forced cultivation, Ivorian cotton quality was often as good if not better than this American standard. To ensure uniform quality, the Agricultural Service and Native Provident Societies distributed selected seeds and discouraged farmers from growing local cotton varieties.[1] Cotton growers were also required to sort their cotton before it was ginned to ensure that only the best seeds from non-hybridized plants were collected for the following year's crop. The collapse of the export cotton market and the end of forced labor changed this system of controlled seed multiplication overnight. As one colonial official remarked: "The seed question will remain a delicate issue in 1948 because the regime of total agricultural freedom, unassailable in principal, is incompatible with correct seed multiplication."[2]

In 1947, no cotton seeds were distributed in the Korhogo region by the Agricultural Service. In fact, Jean Lebeuf, the head of the Agricultural Service, recommended that the northern savanna be "eliminated from the cotton zone" because of unfavorable agro-ecological conditions and the "unfavorable disposition of cultivators."

> There has been virtually no interest in cotton production and commercialization in 1947–48. Subject to climatic hazards which compromise one out of every two harvests, having no suitable lands with the exception of a very small part of the southern portions of the Korhogo and Boundiali subdivisions, this region should be definitively eliminated from the cotton zone. It is not clear what variety would produce interesting yields as long as the parasite problem remains so intense.[3]

Lebeuf reported that the parallel market price was double the European commercial house price of 10 F per kilogram. Not surprisingly, most cotton grown in the Korhogo region was traded in small quantities between growers and Jula traders.

In contrast to the northern cotton growing areas, fifty tons of seeds were imported from Togo for distribution in the Bouaké, Katiola, and Séguéla districts. However, opposition to cotton was so strong in these areas that when trucks arrived to deliver the seeds, people refused to unload them.[4] To make matters worse for those concerned with seed quality, farmers had abandoned cotton as a monocrop and were intercropping local varieties in their food crop fields. The effort to ensure uniform cotton quality was further undercut by the Territorial Assembly which slashed the budget of the Service de Conditionnement, forcing it to reduce the number of its cotton inspectors. So little cotton appeared in export markets that year that the Gonfreville textile mill in Bouaké had insufficient supplies to operate. To compete with the buoyant parallel market, Gonfreville offered 21F for a kilogram of cotton. This led to a slight surge in sales in the Bouaké region. The positive response of cotton growers to this higher price did not go unnoticed by merchants and administrators.[5]

The Chamber of Commerce was at the forefront of price reforms throughout this period. It argued that to maintain the interest of peasants in cotton, prices had to be as high if not higher than the 1947 price offered by Gonfreville. The main problem was that the quality of Ivorian cotton had deteriorated and could only be sold in Havre under a low value classification, or "hors conditionnement." While research and development work on new varieties was underway at l'Institut de recherches du coton et des textiles exotiques (IRCT) (Institute for Research in Cotton and Exotic Textiles) experiment stations (see below), the Chamber recommended three marketing reforms to the governor of Côte d'Ivoire. The first was to reduce export taxes and duties on cotton, which would permit merchants to offer higher prices to cotton growers. Governor Pechoux endorsed this reform. For the 1948–49 cotton campaign, he lowered export duties on cotton from 6% to 1%.[6] The Chamber's second recommendation was to eliminate the monopsony position of the Groupement d'Importation et de Répartition (GIRC), which served as an intermediary between cotton buyers in the colonies and metropolitan buyers. The Chamber sought to allow its members to sell directly to metropolitan buyers and thus reduce their costs. In theory, this would allow them to pass on to producers part of their increased profits. Despite the backing of the AOF high commissioner for this proposal,[7] the Ministers of Industrial Production, National Finance and Economy, and of Overseas France ruled in favor of the status quo.[8]

The third recommended marketing reform emerged from discussions at the 1949 cotton conference held in Bouaké. With the apparent backing of merchants, district administrators, and canton chiefs, Governor Pechoux proposed a return to the notorious market fairs (*foires*) of the late 1920s as a means of regulating cotton quality and prices. Despite the "systematic refusal" of the Conseil Général and the Chamber of Agriculture to intervene in cotton markets and the paucity of cotton inspectors since the layoffs at the Service de Conditionnement, Governor Pechoux argued that the fairs would make "producers free to sell their cotton at the place of production."[9] He also envisioned the fairs as a place to monitor, if not regulate, the practices of indigenous traders. Contrary to evidence presented at the cotton conference, Governor Pechoux argued that Jula merchants "will buy at very low prices" in the absence of controlled markets and thus reduce the incentives for peasants to grow cotton. Jean Lebeuf, head of the Agricultural Service of Côte d'Ivoire and a skeptic of cotton development, noted at the Bouaké conference that Jula traders in the Boundiali area were "more exacting than their European counterparts on cotton quality" and paid thirty francs for a kilogram of cotton – nine francs above the going rate.[10] The official attack on Jula merchants must be seen as part of the postwar colonial development discourse that privileged European merchants and textile interests under the guise of cotton quality and price reforms that would supposedly benefit peasant producers.

The cotton fairs did not take place until the 1950/51 harvest. Cotton output nose-dived in 1948/49. Virtually all of it was purchased locally by Jula traders and weavers. However, attractive prices reportedly led both to an increase in area planted in cotton in 1949 and to more cotton marketed.

The fairs proved to be favorable grounds for introducing additional market reforms in the cotton sector. Most important among these innovations was the introduction during the 1951/52 season of a two-tiered price system. To sell cotton, growers had to sort it into two grades: white and clean, and discolored and unclean. Sorting had two beneficial results. First and foremost, it greatly facilitated the process of seed selection and multiplication. The top price was given only for the highest quality seed cotton, whose seeds could be used the following year. By this simple price mechanism, cotton growers became involved in the process of improved seed selection and multiplication. The two-price system also solved the vexing export market problem of non-homogenous cotton. By sorting cotton at harvest time, followed by a second sorting at the gin, peasants would become major actors in ensuring uniform cotton quality for export.

The two-tiered price system nearly collapsed after its first year in operation. Exporters and gin owners complained that some traders duped them by mixing low and high grade cotton and selling it as top grade cotton. As a result of these fraudulent practices, the Gonfreville mill decided that it would only buy a single grade of mixed-quality cotton ("tout venant") during the 1952/53 season.[11] The Chamber of Commerce continued to support the two-tiered system and called for stricter controls at cotton markets. It specifically pushed for a reorganization of the cotton sector in which agents of the new French cotton development company (CFDT) would control the market.

Institutional and organizational reforms

France's continued interest in cotton production in its overseas territories was linked, as in the past, to the French textile industry's vulnerability to fluctuating world market supplies and prices. In the early 1950s, the metropole's cotton fiber needs amounted to 300,000 tons. Its colonies supplied just one-tenth of this demand, most of it coming from French Equatorial Africa. As a result, France was forced to spend 150 billion metropolitan francs each year to import mainly American cotton.[12] A poor cotton harvest in the United States in 1950, and the government's decision to stockpile remaining supplies at the beginning of the Korean War, led to world shortages of cotton in 1951. France was forced to buy lower quality cotton from Syria, Turkey, and Brazil at twice the price of American cotton.[13]

To stem the flow of its foreign exchange earnings overseas, France renewed its efforts in the postwar period to stimulate cotton in its colonies under free market conditions. Merchants and textile industry representatives within Côte d'Ivoire also lobbied vigorously for a new cotton development program that would ensure steady supplies of high-quality cotton at prices that would sustain producer interest. The Gonfreville textile mill, located in Bouaké, led this local coalition for institutional and organizational reforms.[14] The mill's textile fiber needs in 1949 amounted to 1,400 tons of seed cotton – the amount sold to commerce that year in the Bouaké district. Its plans to expand its integrated spinning and weaving complex in the 1950s would require even greater supplies. In concert with cotton gin owners, the Chamber of Commerce, and the Chamber of Agriculture, Gonfreville representatives lobbied senior administrators and Agricultural Service officials to organize a "Cotton Committee" that would address cotton seed selection and distribution, extension, and marketing. Many of the reforms it recommended at the 1949 cotton conference held in Bouaké would be adopted in the 1950s and 1960s. Some of these initiatives were facilitated by the unprecedented investment of French government funds in two companies whose mission was to promote cotton development in its overseas territories: the IRCT and La Compagnie française pour le développement des fibres textiles (CFDT) (the French Company for the Development of Textile Fibers).

IRCT was formed in 1946 with overseas development funds (Fonds d'Aide de Coopération and Fonds d'Investissement pour le Développement Economique et Social or FIDES). In West Africa it established three research stations: M'Pesoba (French Soudan), Anie-Mono (Togo), and Bouaké (Côte d'Ivoire). In Côte d'Ivoire, IRCT's primary mission was to develop a high-quality cotton variety that would thrive in areas characterized by variable rainfall and parasitic diseases. It sought a variety with a minimum fiber length of 22 mm that would do well as an intercrop and perform even better as a monocrop.[15] Between 1946 and 1952 it focused its research and development activities around a variety called *N'Kourala*, which had been discovered in the Sikasso area of Soudan in the early 1920s. The director of IRCT thought *N'Kourala* to be a cross between a local *punctatum* variety and an American variety from the Mississippi delta introduced by the Association Contonnière Coloniale in 1904.[16] It was also experimenting with a number of *barbadense* varieties, as well as *G. peruvianum* and *G. hirsutum*, type *Allen*. Ultimately, it was the *Allen* variety, grown as a monocrop, that became the basis of the dramatic increase in cotton yields in the 1960s (see below).

CFDT, a joint-venture between the French state and private textile interests, was founded in 1949 as La Compagnie des Textiles de l'Union Française (CTUF) (the French Union Textile Company). French government support came from two ministries: Overseas France, which allocated funds from its FIDES program, and the Ministry of National Economy, which provided financing from taxes levied on French textile companies to subsidize the development of cotton in its overseas territories (Fonds d'Encouragement à la Production Textile).[17] Many of these firms also became individual shareholders in the new cotton company. CTUF changed its name to CFDT in 1950 and, after two years of "prospecting trips" in West Africa, began what was to become a long and fruitful collaboration with IRCT, the Agricultural Service of Côte d'Ivoire, and colonial administrators, to promote cotton in the territory.[18]

CFDT was warmly welcomed in Côte d'Ivoire not only because it appeared on the scene with external funding, but also because it promised to address the major issues around which the local textile industry, merchants, and Chamber of Commerce had lobbied the territorial government during the 1940s. The head of the Agricultural Service wrote with great optimism in a circular to district commanders about his agency's collaboration with CFDT in a new cotton development program. "All hopes are possible in Côte d'Ivoire," he wrote with reference to CFDT's plans for cotton development and extension in the territory.[19] In contrast to the tensions reported in the 1940s between local textile interests and the lieutenant governor and director of the Agricultural Service over the future of cotton in the colonial economy, the warm embrace of CFDT suggests a rapprochement in which merchants, local textile firms, and senior administrators agreed to give their support to the French parastatal and its plans to restructure the cotton sector. The optimism of colonial officials was tempered, however, by the prospect of peasant resistance to cotton cultivation in the post-forced labor era. Indeed, the Agricultural Service director stated in his circular, "I am convinced that we must be cautious, patient, and persuasive with our cultivators who . . . have already suffered from the extension of this crop."[20] The creation of the CFDT system came to reflect the interactions of these multiple players under the changing social, economic, and political conditions of the 1950s and 1960s.

IRCT experimented with plant breeding and pest control, and undertook agronomic research aimed at "improving cultivation techniques" of cotton growers.[21] CFDT's mission was to extend and intensify cotton cultivation by introducing the new varieties and cultivation techniques developed by IRCT. Its tasks included seed distribution, replacing local cotton with selected varieties issuing from IRCT experiment stations, expanding the area under cotton, advising growers on the best cultivation techniques to increase cotton quality and yields, and ginning cotton in areas where gins were absent. In defining its mission and particularly the means to implement it, CFDT looked towards French Equatorial Africa, where cotton was cultivated more successfully. There, crop monitors called *boys coton* or *surveillants africains* served as extension agents to diffuse new techniques and varieties.[22] CFDT believed that this close contact between extension agencies and growers was a necessary condition for increasing cotton output. In 1951, CFDT employed twenty-eight African extension agents to monitor its cotton field trials and seed multiplication plots throughout the savanna region. In the area of agricultural education or "propaganda," CFDT noted the use of "cotton catechisms" in Chad and Oubangui-Chari (Central African Republic) by African crop monitors and even school teachers to spread the word about new cultivation techniques and agricultural calendars for cotton. Finally, in the realm of marketing, the director of CFDT in Côte d'Ivoire noted how cotton quality remained high in Chad due both to sorting by peasants in their fields during the cotton harvest and to the two-tiered pricing system which rewarded peasants for sorting.

> In Chad, cotton grown by farmers without administrative pressure, under the surveillance of a cotton monitor (and) directed by a specialized agronomist, is picked according to two qualities. In the cotton fields, men and women harvest with their left hands the yellow or black-stained bolls, and with their right hand the clean ones; depending on its quality, the cotton is put into one of two baskets. The cotton grower (in Côte d'Ivoire), if he sees his interest, will act in a similar manner.[23]

In 1953, IRCT began to experiment with a *barbadense* variety called *Mono*, introduced from another IRCT experiment station in Dahomey (Benin).[24] In 1954 and 1955, the Agricultural Service and CFDT organized field trials in selected villages in the Bouaké region. Pesticides were used for the first time on Ivorian cotton during the *Mono* field trials, and both monocropped and intercropped fields were treated. Fields that were sprayed in 1955 had twice the yields of untreated fields. The following year, cotton yields in treated fields were four times higher than in untreated fields.[25] M. Cluchier, a French consultant on a three-year contract with the Agricultural Service between 1955 and 1958, described his experience in promoting pesticides in the Katiola area north of Bouaké.

The Agricultural Service and IRCT tried to increase cotton yields by spraying pesticides in demonstration plots and in people's cotton fields . . . In 1955 they gave me a tractor, pesticides, and sprayers, and we sprayed the fields of those people who wanted it. Sometimes it took an entire day of talking to a guy before he accepted pesticide applications as a gift. You see, it created problems. There were some people who accepted it but later no longer wanted it. Because . . . in their cotton fields they also grew all their (sauce) ingredients: okra, hot peppers, all their condiments . . . and because of the pesticides, we said, "be careful, you can no longer use those plants," which women were not happy about. That caused problems.[26]

In the Korhogo region, *Mono* was mainly grown as an intercrop without pesticides. The yields of *Mono* intercropped with food crops were significantly higher than existing varieties. In 1955, the average yield of *Mono 54* intercropped with maize in the Katiali region came to 216 kg/ha under the supervision of a CFDT monitor.[27] However, in the absence of crop monitors, the average yield dropped to 100 kg/ha in the region.[28] Nevertheless, the average yield of *Mono* cotton was more than double that of other *barbadense* varieties. Also significant was the 10% increase in gin yields of *Mono* cotton.[29] The increased yield was partly due to the rustic nature of *Mono*; it tolerated intercropping and pests better than previously introduced varieties. The higher yields were also linked to the higher density of plantings. Bernadin Ouattara, an extension agent employed by CFDT in 1957 to promote cotton in the Dikodugu area south of Korhogo, provided details on how the *Mono* program differed from the "traditional" system of intercropping cotton with food crops.

Before the Mono program, they (peasant farmers) planted cotton on just one side of a (yam) mound. For them it was a way of protecting the (other) plants because if they planted it on both sides of a mound it would choke out their food crops. The (*Mono*) cotton plants grew tall and their leaves were large . . . To obtain higher yields, our bosses told us to make them plant both sides of each mound . . . They planted yams three months before cotton . . . We saw that growing cotton on both sides of the mound did not prevent yams from growing . . . They eventually accepted it but it was difficult. You had to have a hard-nosed philosophy with planters . . . The peasant was in fact very suspicious because he had truly suffered while growing cotton during the colonial period. When we arrived, some people said, "They are here to trick us; this is forced labor that they want to reintroduce." We told them that it (forced labor) was not coming back, but they were very suspicious just the same.[30]

Although *Mono* was selected to replace all other cotton varieties in 1956[31], CFDT was not very pleased with its performance. Cluchier noted that:

Mono presented enormous problems because if it was planted on good soils and it rained a lot, it grew as high as this ceiling (pointing to the ceiling above him in his home). In some years, because the plant cover was so dense, IRCT had to pull out every other row of plants just to spray the field. And often, one had to spray while walking backwards because you couldn't walk forward in such a field . . . Moreover, the capsules were small, of poor quality . . . and the investments made in spraying were high relative to the yield . . . It was at this point that IRCT began its *Allen* field trials.[32]

While IRCT, CFDT, and the Agricultural Service carried out their experiments in the Bouaké region, cotton continued to be grown as a secondary crop in peasant fields in the Korhogo region. In 1962 cotton was intercropped on 21% of the total cultivated area. Just 1% of this area was monocropped. The dominance of food crops within the Senufo agricultural system was partly linked to the demand for food crops in urban areas and the forest region. Indeed, what is most striking about the postwar period is the ongoing role of the northern savanna as a labor and grain reserve for the plantation economy of the forest region. This role of the north was closely tied to the emergence in 1946 of an organized group of African coffee and cocoa growers, for whom the possibilities of accumulation depended on their access to an abundant immigrant labor force.

Migrant labor and the "climate of freedom"

At the Dakar Economic Conference of 1945, the decision made for Côte d'Ivoire to specialize in coffee, cocoa, and rubber production marked the beginning of the forest region receiving a preponderant share of economic development funds.[33] One result of this uneven regional development was the sharp inequalities in regional incomes. In 1956, the average annual income of families in the north was 15,000 FCFA versus 100,000 FCFA per family in the south. The local textile industry and Chamber of Commerce would ultimately cite these differences in their lobbying efforts for greater government investments in the cotton-growing areas. During the 1950s and 1960s, northern farmers had few economic choices to improve their living standards. Migration to cities and the forest region was the most common means through which northerners, especially young men, sought to mitigate their relatively depressed economic situation.

The end of forced labor in 1946 led to a major reorganization of labor recruitment in Côte d'Ivoire. To sustain the seasonal migration of northern labor to the forest region, coffee and cocoa planters, in conjunction with the state, devised new recruiting methods and offered new incentives to attract labor southwards. Under the instigation of Governor Latrille, the postwar colonial state facilitated the movement of plantation workers by offering free transportation by truck and by train from Ouagadougou, Korhogo, and Bouaké to the forest region.[34] By raising daily salaries from 3F.50 to 20 F, and by promoting sharecropping with one-third of the harvest going to workers, African planters hoped to direct the flow of migrant labor to their own plantations.[35] Such incentives were especially aimed at the Mossi of Upper Volta (Burkina Faso) who for years had been bypassing Côte d'Ivoire to work for higher wages and in better living conditions in the coffee and cocoa plantations of the Gold Coast (Ghana).[36]

To recruit labor, planters and professional labor contractors traveled throughout the north trying to attract young men south with alluring descriptions of plantation work. In 1950, when the colonial government stopped subsidizing the transport of migrant laborers, both African and European employers in Côte d'Ivoire founded the Interprofessional Syndicate for the Conveyance of Labor (SIAMO). The association was established to continue subsidizing the free transport of laborers by rail and by road.[37] SIAMO centers were established primarily to orient and coordinate the movement of migrant workers to the forest region. They also sought to eliminate the deceptive practices of professional labor contractors by standardizing recruitment procedures.

The importance of migrant labor to the forest region economy is suggested by Samir Amin, who notes the extraordinary demographic changes that occurred in Côte d'Ivoire between 1950–65. In that period, the number of African immigrants working in the country rose sharply from 100,000 in 1950 to 950,000 in 1965. Two-thirds of the immigrants settled in the forest region, where they comprised 22% of the rural population in 1965 in comparison to 2.5% in 1950. For the most part composed of young men from Upper Volta and Mali, this migrant labor force represented 35% of the rural *male* population in the south in 1965.

The influx of migrant labor into southern Côte d'Ivoire resulted in a major shift in the regional distribution of the country's rural population. In 1950 the north accounted for 34% of Côte d'Ivoire's rural population. By 1965 this percentage had dropped to 27%.[38] The emigration of northerners to southern plantations also contributed to this change in the distribution of the country's population. Although Mossi immigrants from neighboring Upper Volta constituted the major portion of the plantation work force, the part of the "korhogos" from northern Côte d'Ivoire grew in importance during this period.

In the decade following the Houphouet-Boigny Law of 1946 that ended forced labor, there was a notable decline in emigration from the Korhogo region. Despite the greater incentives offered to the "korhogos" (higher wages, better working conditions, and free transportation to the work site) and the recruiting efforts made by southern planters, labor contractors and SIAMO, it appears that few men left their villages to work in the south during the postwar period.[39]

The results of a retrospective migration study, undertaken in Katiali in 1982 with a random sample of 35 households, verify this lack of interest of northerners towards forest region work in the late 1940s and early 1950s. Just a third of the migrations outside of the village between 1946 and 1965 occurred between 1946 and 1954. All of these migrants were Jula men. Some 60% of these migrants worked in the north, with more than half engaged in commerce and weaving. Only 25% of these migrants worked in the south. Half of the 25% who went to the forest region worked in the coffee and cocoa plantations of relatives, while the other half started their own plantations.

The initial reluctance of northerners to emigrate changed by the late 1950s. As shown in Fig. 4.1, the number of migrants working outside the Korhogo subdivision *doubled* between 1957 and 1963. A number of factors help to explain this renewal of labor migration from the region. First, following the capitulation of Houphouet-Boigny and the PDCI–RDA to French colonial political and economic interests during the early 1950s, the discriminatory agricultural and labor policies initiated by the Pechoux administration in favor of French coffee and cocoa growers were gradually lifted. Encouraged by more favorable policies and higher world prices for their crops, African planters expanded their plantations in the forest region. By the mid-1950s, the area in coffee and cocoa trees under African ownership had significantly expanded.[40] Most importantly, this expansion depended upon migrant labor from the northern savanna regions of Côte d'Ivoire, Mali, and Upper Volta.

A second factor behind the increase in labor migration in the mid-1950s concerned the delayed effects of the end of forced labor on the liberation of domestic slaves. Although slave raiding and trading ended with the advent of colonial rule in northern Côte d'Ivoire, for reasons related to political and economic stability, the colonial state did not liberate house (hut) captives (French: *captifs de case*; Senufo: *koulon pigué*; Jula: *woroso*).[41] House captives were the descendants of trade or war captives who were born into a

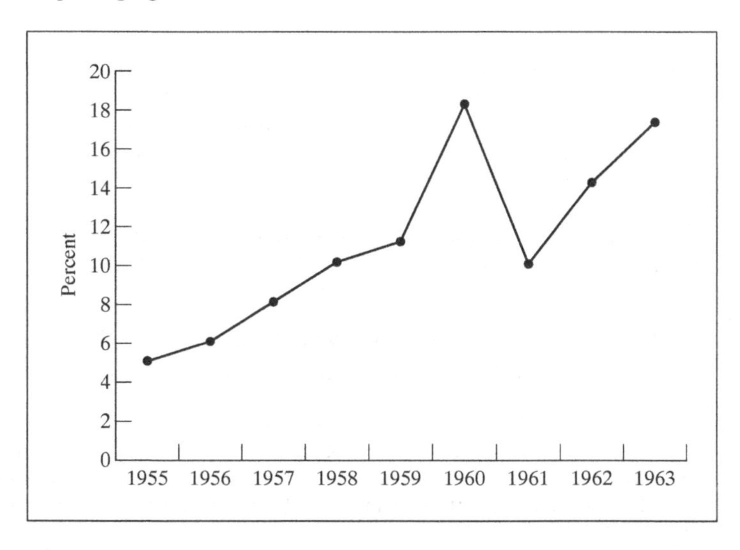

Fig. 4.1 Emigrants from the Korhogo region, 1950s (*Source:* SEDES, *Rapport Démographique*, p. 78)

slave family. In return for their labor and good will, house captives received food, housing, clothes, and spouses. They also had the right to cultivate individual fields and engage in commerce, although they were commonly required to give a percentage (no more than one-fifth) of their earnings to their owners. During the era of forced labor, household heads commonly sent house captives to work on public works projects or in the forest region rather than send one of their own children.[42] According to informants in Katiali, prior to 1946, slaves reportedly worked diligently to avoid being sent to the forest region year after year. It was common that industrious slaves remained in the village while their lazier counterparts went south. With the threat of forced labor no longer looming over them, slaves reportedly did not work as hard. Moreover, they were considered a liability to household heads, who had to feed and clothe their dependants.[43] Many of these liberated slaves joined the migrant labor force as a means of obtaining some economic security. Indeed, 39% of the migrations reported from Katiali between 1946 and 1965 were undertaken by individuals identified by informants to have been former domestic slaves.[44]

Third, young men wishing to purchase consumer goods like bicycles left their villages, often in the middle of the night, to work in the south. Kaléléna Silué, the chief of Katiali between 1983–96, described the clandestine nature of this emigration.

I went south to work in Ferké loading trains to buy a bicycle. Leaving was the only way of earning money to buy a bike. I also worked in the Bouaké area mounding fields. In the village, you had to work for the elders and couldn't earn enough . . . The elders were not happy about us leaving. We had to do it secretively in the early morning hours when everyone was asleep. We would tell someone that we were leaving so that in the morning when the elders found us missing, they knew where we were. We would often go to Pitiengomon (a neighboring village) where Baoulé and Bété planters sent trucks to transport workers south . . . When we returned to the village with our bicycles, everyone was okay. The elders were happy that we returned and we were happy to have bikes to ride to the fields.[45]

Katienen'golo Silué recalled that many young men went to Korhogo where they met up with "bosses" who took them south to work in coffee plantations.

We would be gone for 6 months, sometimes a year, and come back with a bicycle. Everyone saw that you could earn some money if you left the village so people left to work for a bike. It wasn't like that during forced labor; if the workers returned, they were in poor health.[46]

A household budget study undertaken in the Korhogo region in 1962 shows that rural households had developed a need for a cash income as indicated by their involvement in the buying and selling of commodities. The largest categories of *per capita* monetary expenses were food (38%), drinks, tobacco, stimulants (19%) and clothing (11%). Crop and livestock sales accounted for more than three-quarters of *per capita* monetary incomes.[47]

The tensions and conflicts between young men and elders around labor migration led to far-reaching changes in the socio-cultural organization of agricultural production. This was most commonly expressed in the break up of large units of production and the collective fields (*segnon* or *segbo*) of the Senufo. During the pre-colonial and early colonial periods, agricultural production relations among the Senufo were organized at the lineage level. Each lineage resided in the same residential area (*katiolo*) within a village. The oldest male in the lineage held the title of *katiolofolo*, or residential quarter chief. The members of each *katiolo* worked in a collective field (*segnon*) four days a week (*Noupka*, *Tori*, *Kali* and *Tiefonon*).[48] During the remaining two days of the week (*Koundiali* and *Kong*), the residents of these quarters either worked in the fields of the heads of conjugal families (*kagonbile*) or, less frequently, in individual fields (*kagon*).[49]

Young men commonly worked for the *katiolofolo* until they became initiates (*tyolobélé*) in the final stage of the *poro* society. During this six-year initiation period, the *tyolo* also worked in the village chief's fields three days a week (*Noupka*, *Tori* and *Kali*). The remaining three days were usually devoted to work in the fields of lineage heads.

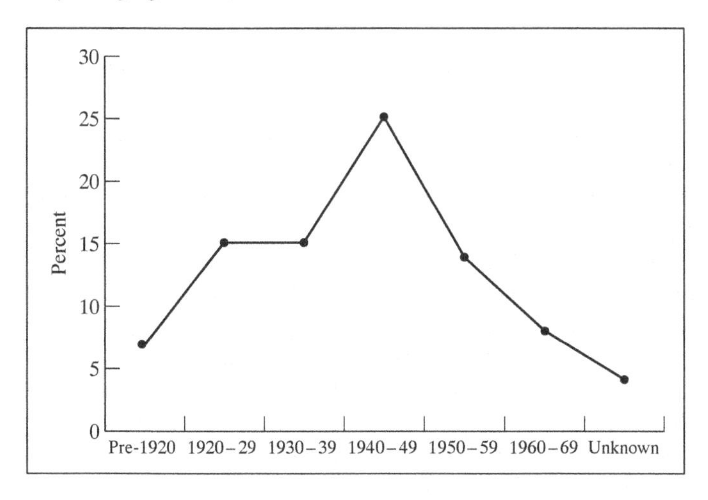

Fig. 4.2 Break-up of Senufo collective fields by date (*Source:* SEDES, *Rapport Sociologique*, p. 26, N = 88)

Marriages were usually arranged either by the village chief or by the lineage head. If a young man worked in the village chief's collective field or in a particular quarter's *segnon*, he was rewarded for his labor with a *segnontio* – "a wife of the collective field." For young men who spent most of their time working outside a *segnon*, a bride could be obtained in any one of three principal ways. A maternal uncle might reward his nephew with a "wife of the lineage," or *nerbatio*. A father might pass on to his son a woman given to him from another lineage – an arrangement known as *tofotio* or "son's wife." Finally, if a young man worked with ten to twenty of his friends one day per year for fifteen to twenty years in the fields of another lineage, he might receive a "wife for doing a good deed," or *katienetio*.[50]

All of the institutions related to agricultural production and social reproduction experienced severe strains during the era of forced labor. The *segnon*, in particular, became the locus of strained relationships between quarter chiefs and household heads, and between young men and elders. Many of these collective fields broke up prior to 1945 (Fig. 4.2).[51] The "climate of freedom" prevailing in the Korhogo region at the end of forced labor was, according to Louis Roussel, "often utilized in villages to suppress an institution experienced for quite some time as unadaptable but which could not be easily abolished."[52] Conflicts between village youth and elders over labor control became so acute that even *poro* societies began to dissolve. In some communities, the *poro* was completely abolished. In most cases *poro* initiations and responsibilities were significantly modified. By the mid-1940s, for example, the period during which *poro* initiates (*tyolobélé*) in Katiali had to remain secluded in the sacred grove was shortened from three months to one month. Nevertheless, the *tyolobélé* continued to work two days a week in the fields of the village chief.

Religious ferment also played a part in the dissolution of initiation societies among the Senufo and Jula. In 1948 Sekou Sangaré, a Moslem marabout from Guinée, began proselytizing in the north. Three years later, Senufo elders of the *Pkeenébélé* subgroup living in a residential quarter of Katiali burned their *poro* society's sacred masks and converted to Islam. The Jula of Katiali initially curtailed their secret society's (*lon*) activities around 1946. However, the *lon* was revived a few years later when a prominent Jula man died in Korhogo. In 1949 another *lon* initiation group entered the Jula sacred grove but never graduated. Jula elders decided to dissolve the *lon* in 1955. These widespread changes in religious practices were perceived by the colonial administration as a threat to political stability. Governor Pechoux, in particular, considered Sekou Sangaré a threat to the *status quo* at a time when French settlers were struggling to defend their privileges.[53]

> The activity of this *marabout* is dangerous from both the religious and social standpoints; the brutal disappearance of the ancient fetish core and rapid expansion of an intolerant Islam on the one hand, and the menace of the destruction of the ritual societies comprising the armature of the hierarchical organization of the autochthonous society on the other hand. It is not surprising that in these conditions that the RDA (Rassemblement Democratique Africain) resolutely supports the advance of Sekou Sangaré.[54]

Colonial administrators were particularly concerned by the loss in prestige of the chiefs they relied upon to support and implement their policies. The authority of loyalist indigenous authorities began to decline with the end of forced labor in 1946 when chiefs lost their coercive power over their population to exact forced cultivation and forced labor. Their status further deteriorated when the RDA, in its efforts to win the support of the peasantry for its anti-colonial political platform, attempted to discredit those chiefs who had collaborated with the French. In their attempts to preserve their position, many chiefs resorted to sorcery to instill fear in and command respect from the population. Indeed, a "recrudescence in fetishism" became apparent to colonial authorities. As Governor Pechoux noted in 1948 in his annual political report to the governor general of AOF:

> It is curious to state at this time that the African chiefs who still have some authority only maintain it as a result of their belonging to fetish sects and to their prominence within these sects; one can easily understand that they are resorting to their religious powers to compensate for the loss of prestige and authority suffered on the administrative level.[55]

A second manifestation of the political-economic crisis and religious ferment at this time was the brief but widely popular "cult of the horn" or "Massa Cult." Introduced into the Korhogo region in 1952, the Massa Cult appeared to be a reaction against the resurgence in sorcery and fetishism. To protect oneself from the magic of sorcerers, one customarily searched for remedies prepared by specialists, whose concoctions were "sometimes complicated and almost always costly."[56] According to Holas, one of the major appeals of the Massa Cult was the perceived power of its deity to protect adherents from all types of sorcery – and at a much lower price. This functional attribute of the Massa Cult, however, is easy to overemphasize in explaining the widespread influence of the cult among the Senufo of the Korhogo region. As Holas commented in a paternalistic tone:

> It would be perfectly unjust to deny to the existence of (the) Massa (cult) more profound aspirations coming from a sincere religious need. During this period, the Senufo, conscious of it or not, were searching for a way towards uniting dispersed spiritual sentiments, pushed in this direction by the progressive relaxation of the *poro* – the rhythm of initiations becoming too slow for the rhythm of the modern world of which he is henceforth called upon to be a part.[57]

The changes in the "rhythm of initiations" to which Holas alludes were far reaching in the realm of labor relations between village youth and elders. The most significant change in the *poro* society during this period was the age at which young men were allowed to become initiated. In the past, men who had reached the *tyolo* stage were usually in their early to mid-thirties. By 1962, the age of initiation was closer to twenty than thirty.[58] In the context of Senufo society, this reduction by ten years in the period of "adolescence" had profound implications for the relations of production and exchange between young men and elders. Since graduation from the *poro* marked the rite of passage to manhood and economic independence for young men, the lowering of the *poro* age meant that elders lost control over ten years of labor from this particularly vigorous age-group. This loss of labor was most severely felt by the household heads for whom the young men worked until graduation from the *poro*. Lineage heads and village chiefs in whose fields the village youth worked two to four days a week until their *poro* graduation also saw their labor pool greatly diminished. Such changes in the rhythm of labor exchanges heightened tensions between elders and village youth.

The elders' frustration and bitterness is easily understandable. As youth, they had worked hard for their elders, all the while believing their efforts would be fully compensated when they became elders. Now that it was "their turn" to benefit from the labor of young men in the village, the youth were rebelling. One Senufo elder eloquently expressed the shift in power relations between juniors and seniors when he declared that "it is the youth who command elders today."[59]

The transformation of production relations accompanying the changes in the *poro* and the collapse of Senufo collective fields (*segnon*) meant that new marriage arrangements had to evolve in order to replace *segnontio* marriages. Not surprisingly, the two marriage types that gradually replaced the *segnontio* model, the *kékourougou* and *katienetio*, were arrangements that allowed household heads to recapture a portion of the labor lost in the social and cultural change taking place at this time. The *kékourougou* marriage involved labor prestations to the fiancée's family. The equivalent of twenty-five man days per year was required, beginning many years before the actual marriage and *continuing indefinitely after the marriage*. Moreover, the wife continued to reside in the compound of her brother or maternal uncle. If the post-nuptial labor gifts were deemed unsatisfactory, the wife's brother or uncle could terminate the marriage. Under the *kékourougou* arrangement the husband, who might live in another village, visited his wife a few times a week for the evening meal and to spend the night. The children issuing from this marriage belonged to the brother or the maternal uncle of the wife, although the husband could request that a daughter be given to him.[60]

Under the *katienetio* marriage arrangement, a young man typically worked in the fields of his prospective in-laws with ten to twenty of his friends once a year for up to twenty years. In many cases, young men began to work for future in-laws *before* a female child was even born into the family. Under this arrangement, the wife lived in her husband's compound. However, the eldest son and daughter of this marriage belonged to the maternal family of the wife. The remaining children stayed with the husband. Two variants of the *katienetio* model were the *tofotio* and *nerbatio* marriages. If a son showed respect and remained attached to his father's or maternal uncle's production unit, he might be rewarded with a wife who had been betrothed either to his father (*tofotio*) or uncle (*nerbatio*).[61]

In all of these marriage arrangements, the fact remained that "the principal trump-card of the elders is that they still control the circulation of women."[62] That is, young women were used by family heads as a means of obtaining agricultural labor from village youth. This dominance of *family heads* over young men was in contrast to former patterns of control when quarter chiefs and village chiefs arranged marriages under the *segnontio* system. The control that individual family and lineage heads now exerted over young women reflected the greater autonomy of these households from *segnon* production relations.

Another indication of how elders tried to recapture the labor of young men through their control of marriage arrangements was in the length of time men had to work to obtain a wife. Although the *poro* age had been reduced, the marriage age for men still remained very late. In 1962, 50% of the Senufo men in rural areas between the ages of 35 and 39 were still single! At the same time, 70% of the women age 20 and over were already married. It was common, as Roussel noted, for sexagenarians to have two or three wives while a young man of thirty was still single. This "monopolization" of young women by elders heightened tensions between generations even further.[63]

Despite the fact that village elders held the "trump-card" over village youth, young men held the option of migrating to the "Basse Côte" to work as wage laborers in the burgeoning coffee and cocoa economy. In many cases, compromises were made between elders and potential migrants. For example, some elders offered individual fields (*kagon*) to young men who could work in them one to two days a week. Most importantly for the eventual expansion of cotton, the owners of these fields had the right to dispose of their crops in any way they wished.

The data provided by SEDES for the Korhogo subdivision indicate that in 1962–63, 2,200 persons were absent from the region. Half of this number were considered "definitely" absent, with the remaining 1,100 classified as "temporarily" absent. Temporary migrants were absent on average for 20 months. The strained relations between village youth and elders are further illustrated by the fact that 25% of the men in the 20–24 age group were absent in 1962.

Whether emigration was motivated by young men's desire for economic freedom from elders or by slaves desire for freedom from masters, uneven regional development was a basic condition stimulating much of this mobility. Personal incomes in the north, based on commercialized agricultural production, were *seven times lower* than average incomes in the forest region.[64] The absolute priority given to the development of the south, particularly in terms of rural development funds allocated between 1946 and 1965, left many northerners with little choice except to work outside their regions to meet cash needs and generally to improve their livelihoods. It was in these social and economic circumstances that the new cotton program was introduced by CFDT in the 1960s, with the coercive backing of the newly independent state.

The CFDT system

While some colonial officials saw the savanna region as a labor reserve for the plantation economy of the forest region and others thought it wiser to promote more profitable crops such as rice and yams, cotton traders, textile firms, and the Chamber of Commerce lobbied strongly for a restructuring of the cotton sector. The latter argued for technical, organizational, and fiscal reforms that would result in higher-yielding cotton varieties and more attractive producer prices. Their efforts gained ground in 1955 when the General Government created a price stabilization fund for cotton (Caisse de stabilisation des prix du coton en AOF).[65] This institutional reform addressed the perennial problem of fluctuating cotton prices and thus introduced an important measure of producer price stability. By far, the most significant breakthrough occurred at IRCT experimental farms, where a new high-yielding hirsutum cotton variety (*Allen* 333) was selected after many years of research and development. *Allen* had originally been introduced into Côte d'Ivoire from Nigeria in 1927.[66] However, after a few years of field trials conducted at the cotton experiment station in Bouaké, *Allen* was eliminated from further trials, principally because of its extreme sensitivity to parasitic diseases.[67] During the 1950s *Allen* was cultivated in the Soudan and Upper Volta, and the Gonfreville mill imported *Allen* fiber from these countries to mix with *Mono* cotton whose fiber lacked the fineness to make high-quality cloth.[68]

The reintroduction of *Allen* in 1959 was made possible by the development of new pesticides (DDT, Enrin, and Lindane) which were applied in monocropped fields. In addition to the new pesticides and high-yielding cotton variety, the "cotton package" promoted by IRCT and CFDT also included fertilizers and a strict cropping calendar in which the planting date was of paramount importance. In 1960, the year of Ivorian independence, CFDT trained 600 cotton extension agents to promote the new variety. It worked closely with the Agricultural Service, which also trained extension agents for the *Allen* program.[69] After three years of very promising field trials,[70] CFDT established an agreement in 1962 with the Côte d'Ivoire government to promote the expansion of *Allen* throughout the central and northern savannas. CFDT not only assumed responsibility for the cotton extension program but also received the privilege to control all aspects of cotton ginning and marketing.[71] The merchant and textile industry-led "Cotton Committee" of the 1940s had long promoted the concept of exclusive buying and ginning zones, ostensibly to ensure seed quality and guaranteed prices.[72] This monopsony over the cotton market gave CFDT an enormous strategic advantage in restructuring the colony's cotton sector during the first decade of independence. First, it effectively eliminated Jula traders from competing with CFDT. As the previous chapter showed, the parallel cotton market dominated by Jula merchants provided cotton growers with some leverage in the sphere of cotton marketing during the colonial period. Peasant farmers would see this limited autonomy erode as CFDT consolidated its monopsonistic powers during the early 1960s. Second, CFDT's exclusive market control allowed the cotton company to erect a vertically integrated research-production-marketing institutional structure which is credited by many as the source of the cotton success story in West Africa.[73] Similar to contract farming, CFDT's market control allowed the company to intervene at the point of production in unprecedented ways.[74] For example, at the heart of the CFDT system was the company's ability to advance inputs on credit to growers. Extension agents could now simply deduct from growers' earnings the sum owed to the company. This simple and efficient means of credit repayment facilitated the rapid introduction of new agricultural technologies. More important still was the fixing of producer prices at relatively attractive levels, which allowed cotton growers to purchase these new technologies as well as consumer goods.

The colonial-era marketing boards (Caisses de Soutien) became Ivorian institutions following political independence. The Côte d'Ivoire government assumed control of the marketing board in 1962, and in 1964 transformed it into a publicly owned corporation called La Caisse de Stabilisation et de Soutien des Prix et des Productions Agricoles (CSSPPA or Caistab) (the Côte d'Ivoire Price Stabilization Board).[75] That same year, Caistab began to protect cotton prices from fluctuations in the world market by offering producers 33.50 FCFA/kg of seed cotton. In reality, the support fund functioned as a tax on producers, since world prices were on average higher than the support price.[76] The originality of the Ivorian institution was its partnership with private companies who served as intermediaries in the marketing of export crops.[77] In the case of cotton, CFDT received the right to be the sole buyer of peasant-grown cotton and the sole seller. An arrangement was also made between the Caistab and CFDT by which the cotton company would be reimbursed for more than a dozen marketing and production "costs" incurred throughout the year.[78] If the world market price for lint cotton was greater than CFDT's operating costs, the cotton company was required to transfer the surplus to Caistab. These funds were then used, in principle, to support producer prices in subsequent years when world market prices were low. In fact, when world market prices fell in the late 1980s and early 1990s, CIDT reduced producer prices (see chapter 6).

As Robert Bates has argued, marketing agencies (or marketing boards) like Caistab served a much larger purpose than price stabilization – if and when they did play this role. They also became the primary means by which newly independent African governments obtained scarce foreign exchange to invest in social and industrial development programs. Yet the "revenue imperative" faced by African governments encompassed more than economic development.[79] Funds were also needed for political purposes, to reward loyal cadres and to woo influential elites to support the ruling party's policies. During the 1970s and 1980s, Côte d'Ivoire's marketing board served this dual role. It regularly diverted funds from Caistab to its *Budget spécial d'investissement et d'équipement* (BSIE) (Capital Investment Budget) to finance the expansion of the industrial sector as well as the large number of public corporations (*les sociétés d'état*) and parastatals (*les sociétés d'économie mixte*) whose managing directors were linked to networks of political patronage.[80] Although 60% of the Caistab's revenues were supposed to be held in reserve to support producer prices, in fact most of it was transferred to the Ivorian treasury to finance large-scale investment projects in both the industrial and agricultural sectors.[81] In 1978 Caistab transfers accounted for 79% of the Ivorian government's investment budget.[82]

For the marketing board to fulfill these political-economic functions, a monopsony was both a necessary and efficient mechanism to transfer funds from the agricultural sector to the state. However, in contrast to marketing agencies in Anglophone countries where public organizations assume this role, the Ivorian marketing board worked through private and semi-public agents like CFDT to purchase and sell agricultural commodities.[83] In this way, the creation of the "CFDT system" was linked to the revenue imperative of the Ivorian government as much as to solving the "problem" of the parallel cotton market. The government's alliance with CFDT signaled an important change in the structure of collaboration between the state and private commercial interests at independence. In contrast to the colonial period when the state adhered to (at least in principle) a free trade policy, the newly independent state accorded a metropolitan parastatal textile company (CFDT) monopsonistic control in the cotton sector. The new relationship promised to simultaneously serve the political-economic interests of the Ivorian government, the French textile industry, and the French government which was searching for ways to maintain its economic ties with its former colonies.[84] The "CFDT system" was thus born. The one question that remained was whether or not peasant farmers would take up the new cotton package. In 1964, CFDT was ready to test the waters.

5 Making cotton work, 1964–1984

Il fallait casser la collectivité! [We had to break up the collective fields!].

Navigué Soro, former CFDT extension agent, Korhogo, 1995

The elders no longer control the young men of the village. Today, it is the young men who command the elders.

Fongnonbêh Tûo, former head of the *segnon*, Katiali, 1982

Confident that its cotton package would be well received by peasant farmers, CFDT sent out its extension agents to introduce the new cotton program. To their great chagrin, the cotton monitors were not well received by the population. Navigué Soro, a CFDT monitor employed in 1964 to introduce *Allen* cotton in the Guembé and Nafoun areas of the southern Korhogo region, described how he encountered considerable resistance from Senufo farmers to the new cotton package.

When I began to work in the Korhogo region, peasants were growing *Mono* cotton. *Mono* was bought by Mr. Escarré who had a gin here in Korhogo. When CFDT arrived it was Mr. Laroche (CFDT sector chief) who bought *Mono* cotton which did not receive any inputs such as fertilizer and pesticides; yields were between 100 to 150 kilograms per hectare. I was hired to promote the *Allen* variety. We received training at IRCT in Bouaké over a ten- to fifteen-day period and then returned to our areas. I can say that it was not at all easy to penetrate (the local village groups). We had to work with the village communal groups ("les collectivités villageoises"). When we arrived in a village, we said that "the *Mono* that you are currently cultivating is not a high-yielding variety. We have another type of cotton that we would like you to try out." We took the name of the residential quarter chief and told him "we are going to give you a small field." They said, "All right, but isn't this the return of slavery?" (alluding to forced cotton).

When they agreed to grow cotton, we measured a field. We used 100-meter ropes, which we extended two times; they found this to be too large a field. So we gave them one hectare, if the village was comprised of one quarter. If there were five residential quarters, we gave them at least five hectares . . . Sometimes it was very difficult. If there were fifteen to twenty young men in the village, to grow cotton the correct way, we were forced to divide a hectare into ten, five, or six rows for which each person was responsible for a part. To spread fertilizer, each person looked after his part and the same with insecticides. At harvest, the crop was delivered to the quarter chief, and after we deducted the costs of the inputs, the money was immediately given to the quarter chief. And he did with it what he wanted.

On some occasions, because people did not tend their fields properly, we had to go to see the subprefect to put pressure on them. People were not at all happy about this . . . The authorities came and took them away and held them for two or three days, even four days at the subprefecture. And while they were there, those who remained in the village had to work in the field. Once the work was completed, the subprefect was notified and the persons being held were set free.[1]

One of the principal reasons for peasant resistance to cotton was the threat it posed to food production. Unlike *Mono*, which was intercropped with food crops, *Allen* cotton had to be cultivated separately as a sole crop. Knowing that peasants were reluctant to give up the *Mono* variety, CIDT resorted to subterfuge to eliminate it from farmers' fields. In collaboration with the privately owned cotton gin in Korhogo where most of the region's *Mono* cotton was ginned, CFDT arranged for *Mono* seeds to be heated to the point that they could not germinate.

And then, nice and dry, we gave them (peasants) these seeds. They planted them and they didn't grow. That is, we had to destroy *Mono* to promote *Allen*. Yet we still ran into some problems because there were women who sold small quantities of (*Mono*) cotton in local markets. Other women would buy it and remove the seeds (before spinning the lint into cotton thread); these seeds were healthy and they sold them to peasants who wanted to grow (*Mono*) cotton. But we bought cotton in bulk which we sent directly to the mill and there the seeds were removed and killed which we gave back to peasants . . . This is how *Allen* took its place.[2]

Resistance to the new program took a variety of forms; in addition to field neglect, sabotage and sorcery were also practiced. There were also expressions of symbolic discontent, as one beleaguered extension agent recounted:

We extension agents, or monitors as they called us, we experienced things you wouldn't believe. You would arrive in a village, the people would take cotton seeds and pound them to make a sauce to go with some guinea fowl. They would say to us "Since we eat cotton now, you too must eat." When you saw a sauce like that, you couldn't eat it. It was their way of telling us that they didn't want to grow cotton . . . They prepared such a meal a number of times for me. In '64, it was a true conflict around here.[3]

Sabotage took the form of dumping fertilizers in the bush. When CFDT monitors came to visit their fields, peasants would explain that the poor condition of their cotton was proof that *Allen* would not do well there. Sorcery also took its toll on extension agents.

I had a friend, who pushed peasants so hard to grow cotton in the Karakoro area that he was poisoned. They cast a spell on him; maybe you do not believe in spells, but he lost an eye. He lives here in the city; his name is Basile. I knew another, he received an incurable sore; someone cast a spell on him, and up until his death, the sore never healed. He carried that sore with him for years. He tried all sorts of antibiotics and still it wouldn't heal. All that because that agent introduced (*Allen*) cotton to the village. He was seen as bringing hunger to them.[4]

CFDT cotton monitors were not deterred. Each one had their own strategy for persuading people to grow cotton. A common tactic was for a monitor to approach the most recalcitrant person in the village and persuade him to experiment with the new cotton package. Monitors often softened up such persons by talking to them nicely and giving them small gifts like kola nuts and bread. After a while, the person agreed to plant a quarter-hectare of cotton in collaboration with the monitor. Everyone else in the village thought the field was for the CFDT extension agent. But when this "planter friend" was paid after the harvest, people began to talk. They would say how at first he had sworn he would never grow cotton, but obviously he had changed his mind. They concluded that it must be profitable, and some began to ask for a quarter-hectare to cultivate the following season.

One of the greatest challenges facing the cotton program was the continued strength of lineage production units and the institution of the collective field (*segbo* or *segnon*), especially among Senufo subgroups of the southern Korhogo region. From the start of the *Allen* program, extension agents perceived village and quarter chiefs' monopolization of cotton profits as an obstacle to its expansion. In a manner reminiscent of colonial policies promoting individual cotton growing, monitors focused their efforts on young men who were interested in earning money to purchase things like bicycles, shoes, and clothing. Thus, CFDT pursued a policy of breaking up collective fields as a condition for the takeoff of its new cotton program. As recalled by Navigué Soro, former CFDT crop monitor:

The collective field did not further our objectives. Gradually, where we succeeded in loosening the bonds between young men and their elders, *Allen* cotton began to expand . . . It was the result of our breaking up collective fields (*casser la collectivité*) . . . The collective field was there; what we succeeded in doing over time was to persuade individuals that rather than cultivating a (cotton) field for the "boss" to plant a small field on the side, a quarter of a hectare, for themselves. We would say, "since the quarter chief does not give you money, you can sell (cotton from) the quarter hectare for two to four thousand CFA, which you keep for yourself; no one is going to come and bother you." After a while, the quarter chief's (cotton) field was abandoned and it was the young men's fields that prospered. That is how we succeeded introducing *Allen*.[5]

Monitors typically staked out a quarter-hectare field in the proximity of the collective field. This allowed young men to work in their individual fields early in the morning before work began in the collective field. If there was time at the end of the day, they could continue to work in their cotton fields. By locating cotton fields next to the collective field, growers could better monitor their new crop at the same time that it served as a demonstration plot to other members of the collective. After cotton was sold, monitors paid growers in small denomination bills to make it seem like they were earning a pile of money.

Informants in Katiali described how *Allen* was introduced into the area in 1964 as a demonstration crop in the chief's fields. For three years the chief, N'crin'golo Silué, monocropped a quarter hectare of the new variety known locally as *kolochan* or "you need a small sitting stool" – in reference to the relatively low height of the *Allen* variety in contrast to the towering *Mono* variety.[6] In 1967 each residential quarter of the village was required to grow cotton within a large plantation not unlike the *champs de commandant* of the colonial period. Individual household heads had to plant one-half hectare each under the supervision of a cotton monitor. The names of household heads were derived from a "forced census" undertaken to determine the size of production units – especially the number (and names) of married men and women and children between the ages of fifteen and twenty.[7] If someone refused to grow cotton, he was reported to the local authorities by the monitor. On some occasions, the subprefect held meetings in selected villages to promote cotton. At one such meeting in Katiali in February of 1968, the subprefect asked the gathered population who was planning to grow cotton that year and who was not. Bazoumana Diabaté, a Jula weaver and farmer, was the first to respond. He had cultivated cotton the year before, he said, but wasn't planning on growing it again.

> I told him that there was nothing in it for me, and that I had no interest in growing it. The subprefect was angry. He said that it wasn't right to respond to an authority like that. He said that if I said "yes, I was planning to grow it" but did not, then he would understand that I didn't have enough time; but to say "no" the way I did in front of everyone would only ruin the government's objective which was to have everyone grow cotton.[8]

Four other farmers stood behind Bazoumana, declaring their similar intentions not to grow cotton that year. The displeased subprefect ordered them to appear at his office the next day. When the group arrived the following morning, the subprefect ordered them to clear the brush and weeds around his building. They did this for two days. On the third day he called them together and asked: "Are you going to cultivate cotton this year or not?" They responded, "Yes, by force, we are going to do it; otherwise, we would not do it voluntarily." The group left for Katiali and planted cotton that year.

This was not the last time the subprefect used force to promote the new cotton package. Other household heads suffered the same humiliation after their name was given to the subprefect by the crop monitor. The monitor lived in Korhogo but stayed in Katiali for a few days at a time during critical periods in the agricultural calendar. During these visits he noted the names of recalcitrant growers. After they arrived at the subprefecture, the defiant farmers were given hoes and told to weed. When they said they were ready to grow cotton, they were allowed to return to the village. Many people returned to Katiali the same day they left.[9]

Katienen'golo Silué stated that after five years of forced cultivation, people began to grow cotton on a volunteer basis. When asked why there was this turnaround, he said:

> There was no other way to earn money. You couldn't earn anything selling food crops. Five ears of maize sold for 5 FCFA (Francs de la Communauté Financière d'Afrique). When the purchase price of cotton went up to 40 FCFA for (a kilogram of) white cotton, and 30 FCFA for yellow stained cotton, we decided to grow it.[10]

Post-war migrations to the forest region exposed the Senufo to the relative prosperity of the coffee and cocoa sector. The availability of land and the buoyancy of export markets encouraged many to settle in the Basse Côte to establish their own plantations. For those who returned to or never left the north, cotton growing was an alternative means of earning money. The possibility of purchasing more goods, such as bicycles, kerosene lamps, and tools, with cotton earnings was an important stimulus for peasant participation in the *Allen* program.

To accommodate the increasing number of individuals requesting to grow cotton, CFDT cleared and plowed large (10 ha) blocks of land in which individual households could stake out their own parcel. The maximum and minimum size parcels measured by extension agents and farmers were one hectare and one quarter hectare respectively. The policy of cotton blocks was reminiscent of the "commander's fields" of the colonial period, in which residential quarters cultivated distinct parcels within a large block. However, in CFDT's cotton blocks, parcels were cultivated by individual households, in keeping with the cotton company's policy of promoting "individualized" production. Subsistence crops for household consumption were planted in a different and often distant location.

The cotton blocks allowed extension agents to instruct growers how to obtain the highest yields and to supervise specific tasks. For example, in 1967 agents encouraged peasants to construct rows every other meter and to space cotton plants 20 centimeters apart on each row. Growers had to have their seeds in the ground before 30 June in order to receive insecticides for the first spraying. The timing and number of insecticide applications and fertilizer doses were similarly regulated. In 1967, insecticides were sprayed an average of 2.5 times on cotton fields and 160 kg of NPK fertilizer were applied.[11] The doses and the timing of their application changed over time. Both the formula and the dose of fertilizer changed in 1972 when growers spread 200 kg of NPK fertilizer plus 100 kg of ammonium sulfate on their fields.

To CFDT's dismay, the percentage of cotton fields grouped into 10 ha blocks declined from 50% to 43% and then to 30% between 1967 and 1970. Weed infestation and reduced cotton yields led farmers to abandon these blocks and to plant food crops such as sorghum, millet, and peanuts that could benefit from residual fertilizer in the soil. Increasingly, peasants cleared land for cotton adjacent to their food crop fields. The cotton block policy was abandoned in 1974 in favor of individual fields dispersed throughout a region.[12]

The number of cotton growers and the area under cultivation rose rapidly in the 1960s. Between 1963 and 1965, the area increased almost fivefold, rising from 2,518 ha to 11,768. By 1968, there were approximately 68,000 growers cultivating 48,000 ha of sole cotton. Producer prices increased from 30 FCFA/kg in 1960, to 33.50 FCFA/kg in 1964, and to 40 FCFA/kg in 1970. On the one hand, these unprecedented levels of participation and production represented a new chapter in Ivorian cotton history. On the other, this new phase cannot be fully appreciated without an understanding of the changes that took place in agrarian society and economy during the previous fifty years.

First, the sheer numbers of growers enrolling in the *Allen* program reflected a particular social history in which an export-oriented, commodity-producing peasantry had gestated and finally come forth. The collapse of the *katiolo* in Senufoland, the creation of new economic needs, and the emergence of new units of production and consumption during the colonial period were important conditions for the spread of *Allen* cotton in the north. What was different from the colonial period was the way in which foreign capital and the state intervened in the conditions of household production through the CFDT system. The phalanx of extension agents providing technical advice and farming inputs, and the partnership between CFDT and the marketing board (Caistab) in controlling prices and the market, represented a new institutional and organizational form of production and exchange. The coordination and supervision of the production process were similar to contract farming in that a central agency (CFDT) exercised considerable control over various phases of production and marketing. Second, partly as a result of their previous experience with cotton, Senufo and Jula farmers were open to experimenting with new agricultural techniques. They had modified their practices in the past when Agricultural Service agents and district guards imposed new cotton varieties and cultivation techniques in the "commander's field." The appearance of CFDT extension agents with a new cotton package thus fitted into already established practices. What was different were the attractive institutional arrangements and dramatically higher yields which offered more interesting economic opportunities. Cotton growers had also become sensitive to the boom/bust nature of export markets during the colonial period. Guaranteed producer prices, therefore, introduced a certain economic security that had been absent in the past.[13] Third, the cotton development discourse in Côte d'Ivoire at independence contained domestic political, economic, and ideological components that were muted in the interventions of the colonial state. In contrast to colonial cotton policies which were largely formulated in the interest of the metropolitan textile industry, post-colonial cotton policies have had the dual objective of reducing regional income disparities and supplying cotton fiber to the national textile industry. The continuities between the colonial and post-colonial periods are important in explaining the level of involvement and performance of cotton growers participating in the *Allen* program since the mid-1960s. But differences between the two periods are also significant, in that they reflect the very different conditions in which agricultural surpluses were being produced by and extracted from rural producers. The coordination and supervision of the productive process by CFDT and the state contrasted sharply with the colonial regime of requisitions.

The high-yielding variety (HYV) strategy so vigorously pursued by the state at independence seemed logical in light of the failure of colonial agricultural policies to generate sufficient quantities of cotton for export markets. Yet the CFDT-induced model of agricultural intensification represented by the *Allen* program reflected the quality control and yield-maximizing goals of IRCT experiment station agronomists, extension agents, and the state, more than the factor scarcities and interests of peasant farmers. The *Allen* program's bias towards increasing yields per hectare was more appropriate to situations of high population density and relative land scarcity. It was a labor and capital intensive technology that was not easily integrated into the labor-scarce and capital-deficient farming systems of the Korhogo region. In short, what peasant farmers needed was a labor-saving technology rather than the land-saving technology pushed by CFDT.

This example of "policy-led intensification" differs sharply from the induced innovation model.[14] Farmers did not pressure extension agents to develop the high-yielding biological package of the *Allen* program. As we have just seen, CFDT and the state coerced peasants to experiment with the HYV package. Subsequent policies aimed to induce behavioral responses by making cotton growing more economically attractive. During this second phase of directed-innovations, what I call the *seduced innovation* phase, the state offered incentives to cotton growers in the form of administered prices in both input and output markets. The expansion of cotton was not, however, simply linked to these inducements. It also hinged on the ability of farmers to manage successfully the labor bottlenecks in their farming system. The process of breaking labor bottlenecks drove farmers to seek new technologies as well as to modify certain social and institutional arrangements. The resolution of this basic contradiction between the biases of the *Allen* program and farmer preferences within the wider political economy of Ivorian agricultural development policies is the focus of the remainder of this chapter.

The data behind the cotton revolution

The quality of data on African rural economies is notoriously poor.[15] This said, some sources are better than others. Cotton production indicators for West Africa are relatively good because of the monopsony held by parastatal cotton companies in cotton marketing and extension. Other measures – such as the area under cultivation, the number of cotton growers, and therefore yields per hectare – vary in quality. For Côte d'Ivoire, these indicators are relatively reliable for the period 1970–85. The data were collected by extension agents who resided in rural communities and monitored the activities of cotton growers. Up until 1985 they advised cotton growers on new techniques, measured their fields, and distributed fertilizer, insecticides, and seeds.[16] However, following the compression of CIDT's personnel in 1985 in the context of a World Bank structural adjustment program, extension agents were no longer capable of effectively monitoring the activities of farmers. The new extension system promoted by the World Bank and adopted by CIDT ("La quinzaine du moniteur") forced agents to work with more growers (maximum 300) and to cover larger areas. One agent in Katiali remarked:

> Before 1985, there was one monitor for every 100 hectares. After 1985, one monitor was responsible for 428 hectares. The difference (in extension activities) is that prior to 1985, we monitored fields; today we monitor individuals.[17]

Rather than visit individual farmers in their fields, agents now had to meet with groups of farmers on regularly scheduled days in just one location. As a result of these changes, extension agents no longer effectively monitored the nature and extent of cotton growing activities. The devolution of fertilizer distribution to village cooperatives made it even more difficult for agents to record with accuracy the number of cotton growers and the area under cotton. For these reasons, CIDT's cotton data for the post-1985 period should be read as indicative of trends but not of actual practices.

Household survey data, collected in Katiali over the 1980s and 1990s in four different farming systems surveys, suggests that CIDT's data for the post-1985 period seriously underestimates the number of cotton growers and the area under cultivation. By extension, the company exaggerates farmer yields (see below).

Most of the data on "improved" food crop production collected by CIDT are particularly useless. Although these data are meant to record the foodcrop area that has benefited from some aspect of "modernization," such as hybrid maize varieties, fertilizer application, or herbicide, agents simply ask cotton growers how much area they have planted in maize or rice and record this as "improved" on the assumption that farmers involved in cotton growing will also cultivate food crops in an intensive manner (e.g. apply fertilizers, herbicides, etc.). For this reason, CIDT's measures on the intensification of food crops are not used in this study. Only the figures for food crop area that were recorded in the Katiali surveys are used to indicate patterns and trends.

CIDT data for the fifteen-year period 1970–84 suggests that in terms of yields, area under cultivation, and total production of seed cotton, Côte d'Ivoire's cotton development program was an outstanding success. Seed cotton production rose from 30,000 metric tons in 1970/71 to 212,070 tons in 1984/85, representing a sevenfold increase in output (Fig. 1.1). This impressive growth was due in part to a fourfold increase in the area cultivated in cotton and in part to increasing yields.[18] The expansion in area was largely due to a doubling in the number of growers and to a near-doubling in the average area cultivated per grower. Cotton yields fluctuated over this period but were generally above 1,000 kgs/ha. Gin yields also increased from a respectable 39.67% in 1970 to an impressive 41.00% in 1984.

When peasants began to lose interest in cotton, CIDT offered a number of economic incentives to keep them from switching over to alternative cash crops like rice. For example, by 1970 inflation had substantially eroded the purchasing power of cotton growers.[19] With the rising labor demands made by weeding, the additional requirement made by CFDT in 1969 for growers to sort their cotton into two grades, and lastly, late rains in 1969 and 1970, the disadvantages of the cotton program became all too apparent to rural producers. Cotton production fell 23% in 1969 from the 1968 peak of 41,739 tons of seed cotton. Total cotton production dropped by another 7% in 1970 (Fig. 1.1). In its annual reports for 1969 and 1970, CFDT noted that peasants in the Korhogo region were devoting more time to food crops than to cotton.

To make cotton cultivation more attractive to producers, the state substantially raised the purchase price of cotton during the 1970s. A 15% price increase in 1970 was followed by a 75% increase in 1975 and a 14% increase

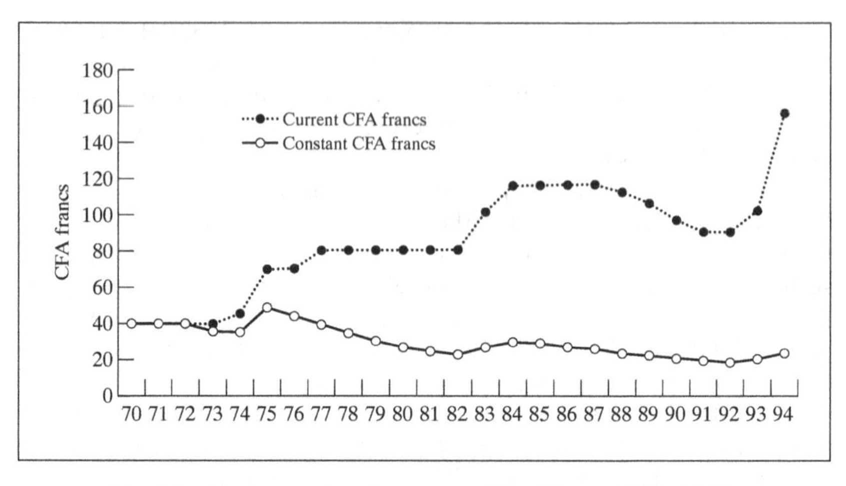

Fig. 5.1 Producer prices for cotton, Côte d'Ivoire, 1970–94 (*Source:* CIDT, *Rapports Annuels*)

in 1977. In current FCFA, first grade cotton prices were double the 1970 producer price (80 FCFA versus 40 FCFA). However, when inflation is taken into consideration, producer prices actually declined by 25% in constant FCFA over the 1970s.[20] Using 1970 as the base year, Fig. 5.1 shows that producer prices in constant FCFA peaked in 1975 (44 FCFA/kg), a level that has not been attained since that year. The downward trend in real prices broke the momentum in peasant interest in cotton. The number of growers dropped by nearly 10% between 1975 and 1976.

To prevent further disaffection from cotton growing, CIDT not only increased producer prices for the 1977 campaign from 70 to 80 FCFA but also gave fertilizer at no cost to cotton growers, a policy which continued until 1983. These incentives clearly worked to stimulate farmer interest in cotton. CIDT data show a 25% increase in the number of cotton growers and a near doubling of the total area under cultivation between 1976 and 1980 (see Fig. 1.1). The expansion of cultivated area was due in part to the increasing number of growers taking up cotton and, in part, to the increase in cotton area per grower. The average cotton area cultivated by each grower increased from 0.99 ha in 1975 to 1.16 ha in 1979 – an 18% increase. Overall, cotton seed production more than doubled over the second half of the 1970s. The intensification of cotton over this first decade is marked by the near universal use of fertilizers and pesticides. In 1980 CIDT reported that (NPK) fertilizers had been applied to 99% of the cotton area and pesticides to 100%.[21] As farmer knowledge and input use increased, cotton yields correspondingly rose from 820 kg/ha in 1970 to 1160 kg/ha in 1979 – a 30% increase.

The introduction of ox-plows played a key role in expanding cotton production. From the start, CFDT's farm mechanization program sought to reduce the competing labor requirements of cotton and food crops by improving the productivity of labor. Animal traction promised to alleviate labor bottlenecks in the planting and weeding phases of the agricultural calendar (see below). In 1970, the year CIDT began to promote animal traction, less than one percent of the cotton area was cultivated by oxen. Ten years later, oxen plowed more than 20% of the cotton area. By 1989, animal traction was used on close to 40% of the cotton area. The expansion of ox-plows in the cotton-growing areas was greatly aided by CIDT's arrangement to serve as an intermediary between cotton growers and the Banque Nationale du Développement Agricole (BNDA) (National Agricultural Development Bank). Cotton growers obtained oxen and plows on credit from BNDA and paid back their loans over a three-year period.[22] Each year at the cotton market, CIDT simply deducted that year's payment from individual cotton grower's earnings and transferred the money to BNDA. Although there is some debate over whether the productivity of labor has increased with this form of mechanization in Côte d'Ivoire,[23] the area under cotton by ox-plow owning households is nearly triple the area cultivated by manual farmers.

Cotton yields and production levels were also affected by the introduction of new hybrid seed varieties and pesticide sprayers. The cotton breeding program based at the IRCT station in Bouaké concentrated on selecting hybrids with superior fiber quality (length, strength) and gin yields. The *Allen* variety was replaced in the early 1970s by a hybrid named HAR444–2, which was itself replaced by a succession of hybrids (see Appendix 1). Gin yields steadily improved, rising from 39.2% in 1967 for *Allen* cotton to 43.5% in 1985 for ISA205. However, the innovation that had the most positive impact on yields was the ultra-low volume (ULV) pesticide sprayer.[24] In contrast to the much heavier, hand-pumped knapsack sprayer which required considerable amounts of water to treat a field (up to 300 liters per hectare),[25] the light-weight, battery-operated ULV sprayer required just one or two liters of water per hectare. This important addition to the cotton package allowed farmers to treat two to six rows of cotton at once, depending on the height of the vegetation. According to CFDT officials, the diffusion of the ULV technique "revolutionized" cotton growing.[26] Its greater efficiency appealed to farmers who were relieved of the burdensome and time-consuming task of transporting large quantities of water to their fields. In Côte d'Ivoire, the new sprayers were given freely to cotton growers along with their completely subsidized (100%) pesticides. Within ten years (1975–84), the ULV sprayer had totally replaced the classic hand pump.[27]

Throughout the 1980s, the cotton sector was affected by a series of International Monetary Fund austerity measures and World Bank structural adjustment programs.[28] The objective of these interventions was to reduce the government's budget deficit and soaring external debt through various policy reforms.[29] The 1983 structural adjustment loan specifically introduced reforms aimed at increasing agricultural production incentives and enhancing the competitiveness of Côte d'Ivoire's principal export crops. As a result of these programs, CIDT eliminated the fertilizer subsidy and raised producer prices. A second producer price increase in 1984 was similarly aimed at increasing production for export. Despite the steady decline in producer prices in constant terms, peasant farmers responded favorably to these back-to-back nominal price hikes. After falling 8% between 1979 and 1982, the number of growers grew by 15% between 1982 and 1984. The other major parameters of cotton also showed signs of improvement between 1979–84: the area per grower rose from 1.25 ha to 1.40 ha while yields, although fluctuating, reached a record level of 1,450 kg/ha in 1984, up from 1,160 kg/ha in 1979. Cotton seed production reached an all-time high in 1984 when 212,070 tons were harvested – a 49% increase over the 1979 level. Despite falling purchase prices in real terms, peasant farmers were still attracted to cotton because it offered access to credit and subsidized inputs. In the absence of alternative cash crops and credit sources, farmers continued to invest in cotton production.

In summary, the impressive rates of agricultural growth between 1970 and 1984 support the claim for an agricultural revolution centered on cotton in Côte d'Ivoire. Over this fifteen-year period, seed cotton production rose sevenfold, increasing at an annual rate of 17%. Cotton yields increased at an average rate of 5% per year. Cotton area grew at an even faster rate of 11% per year. The area growth rate was due both to a more than twofold increase in the number of growers and to an expansion in cultivated area per grower. The latter nearly doubled, rising from 0.77 to 1.44 ha per grower. Yield increases accounted for 38% of the greater output over these fifteen years, and 62% of the growth was due to the expansion in cotton area. This expansion in cotton area was closely tied to the rapid diffusion of ox-plows in the 1970s and 1980s.

Labor bottlenecks and agricultural change

The intensification of peasant cotton at the village level has hinged upon the ability of peasant farmers to resolve the acute problem of labor bottlenecks in an evolving agricultural calendar. The following sections examine the innovative steps taken by the farmers of Katiali to integrate cotton into their labor-constrained farming system. In the first section, I discuss the major environmental and social factors that influence the timing and capacity of households to engage in agricultural work. The following section examines the contours of peasant agriculture with a focus on cropping patterns and peak labor periods. In the third section, I focus on the coping strategies of households of different socio-economic standing to break labor bottlenecks. The final section concludes by arguing that the cotton revolution in Côte d'Ivoire is the outcome of the interactive effects of directed and induced innovations within a wider political economy.

The data upon which this analysis is based were collected from two different samples in Katiali. Labor allocation data were collected on a weekly basis in 1981–2 from a small sample of 26 active workers (14 male, 12 female) residing in 7 households. The data on population, cropping patterns, cultivated area, and farming techniques issue from a sample of 38 households that were systematically surveyed in 1982, 1988, 1992, and 1994. The total number of active workers in these larger samples was 183 in 1982, 308 in 1988, 305 in 1992 and 290 in 1994.

Rainfall is the most important environmental influence on the scheduling of agricultural work. The Meteorological Service of Côte d'Ivoire maintains an observation post in Korhogo where continuous rainfall data have been recorded since 1945. At its meteorological station at Ferkéssédugu, 55 km east of Korhogo, more complete information since 1927 is available covering not only rainfall but also temperature, humidity, evaporation, and winds. Both data sources are used below to describe the effects of seasonality on agricultural work patterns in the Korhogo region. The rainy season[30] lasts between five and six months (May–October), when the area receives more than 80% of its annual precipitation. Most agricultural tasks such as plowing, sowing, weeding, and transplanting, are confined to this period. A more limited number of tasks (land clearing, most harvesting) takes place during the six-month-long dry season (November–April).

For the period 1945–62, average annual rainfall amounted to 1,403 mm for the Korhogo region. The average for the period 1945–80 was 1,104 mm. Figure 5.2 shows the monthly distribution of rainfall for the Korhogo region for the years 1945–80. Of course, these averaged data mask considerable inter-annual variations in monthly rainfall patterns. For the purposes of this study, however, they adequately convey the important point that agricultural activities are constrained by this seasonal pattern of rainfall. Like peasant farmers throughout the tropics, Senufo farmers stagger agricultural tasks, employ non-household labor, and experiment with labor-saving techniques in order to meet their production goals. A comparison of Figures 5.3 and 5.4 illustrates this general correspondence between rainfall patterns and agricultural work in Katiali for 1981–82.

The distribution of rainfall within the rainy season has important

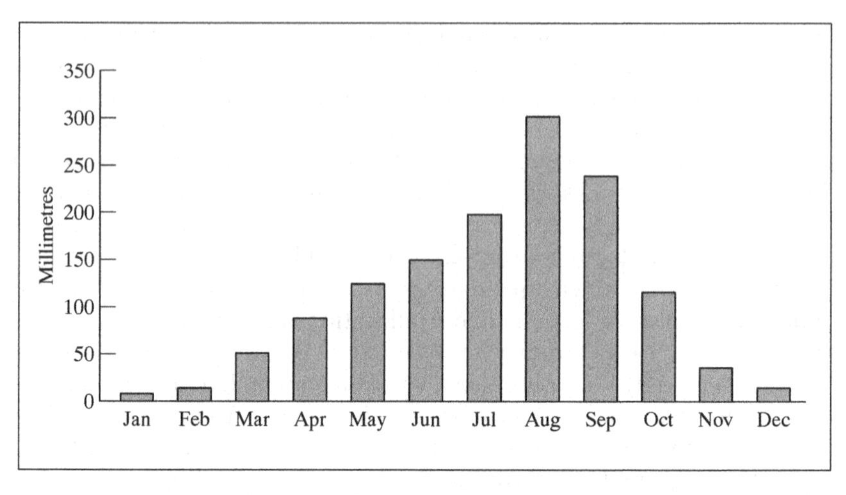

Fig. 5.2 Average monthly rainfall, Korhogo, 1945–80 (*Source:* Dibi, "Regional climate")

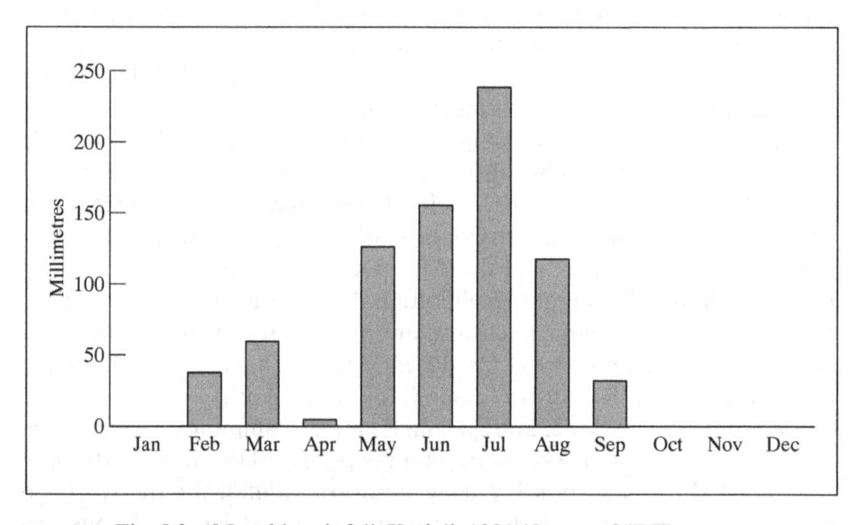

Fig. 5.3 Monthly rainfall, Katiali, 1981 (*Source:* CIDT)

repercussions on the timing of farm work. For example, if rains are late at the beginning of the season, farmers must delay their plowing and planting activities. Such delays exacerbate the planting labor bottleneck (see below) by narrowing the time frame in which fields can be prepared for sowing. Depending on a crop's growing cycle, a short drought in the middle of the rainy season can adversely affect its growth.[31] Too much rain at any period can stall work, accelerate erosion, and adversely affect yields.

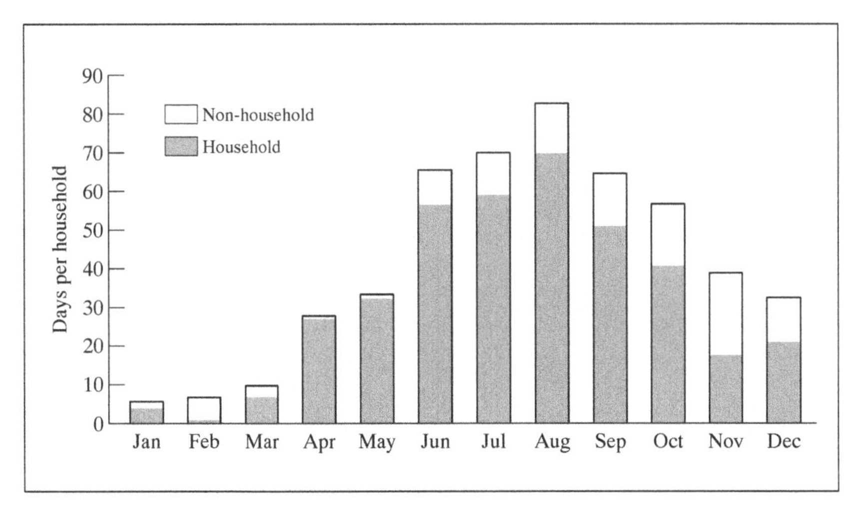

Fig. 5.4 Labor allocation to food crops and cotton, Katiali, 1981 (*Source:* Field data)

Population and labor constraints

The capacity of peasant households to cope with peak labor periods in the agricultural calendar is facilitated or constrained at the most fundamental level by the number of active agricultural workers residing in the household. In villages where the emigration of young men and women is high, peak labor periods are more acutely experienced than in labor-rich communities. The population pyramids for Katiali in 1981–82 and 1991–92 show a large number of young active men as well as women as absent from the village (Fig. 5.5 and Fig. 5.6). This high percentage of young emigrants is not uncommon in the Korhogo region. Fargues estimated that almost half of the men in the 20–29 age group had emigrated from rural areas of the north over the period 1975–80.[32] Emigrants typically go to secondary schools or seek employment in Korhogo, or go to the forest region around Séguela, Bouaflé, and Daloa to work in coffee and cocoa plantations. Their absence intensifies the seasonal pressure on household labor and increases the burden on those who are present, especially women.

The socio-economic differentiation of households in Katiali means that not all households are equally affected by labor shortages. Table 5.1 shows that relatively prosperous households (income group I) have almost twice as many active workers than middle-income households, and more than four times as many actives as low-income households.

A household's active work force is the most important source of on-farm

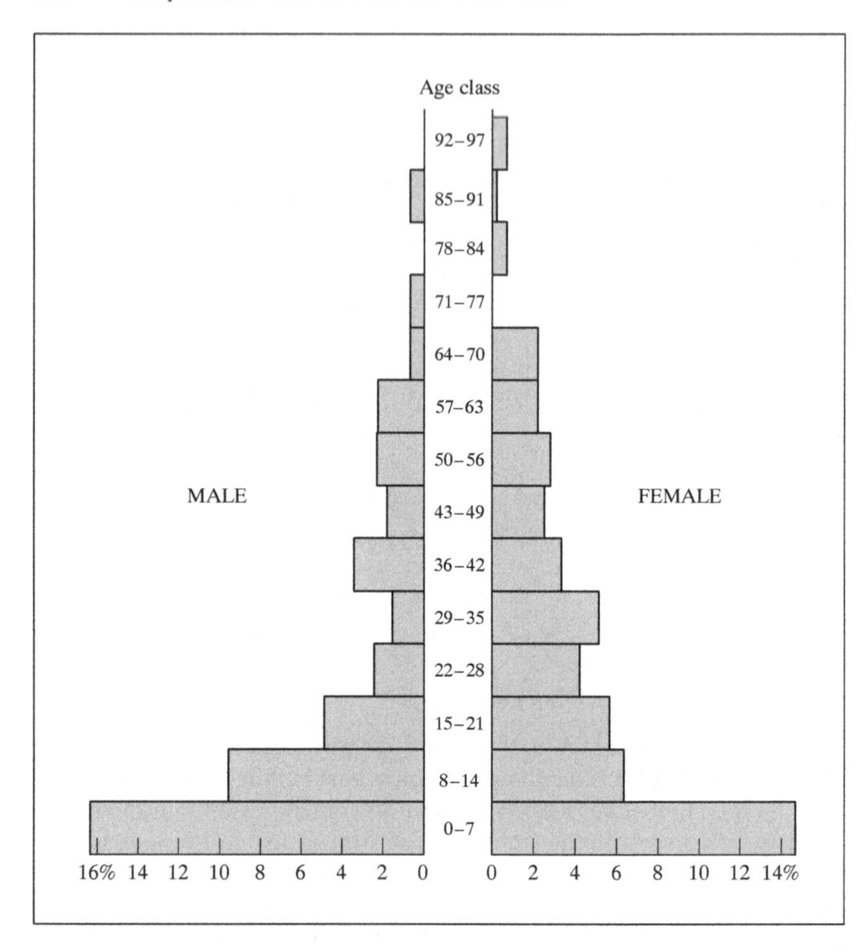

Fig. 5.5 Distribution of population by sex and age class, Katiali, 1981 (*Source:* Field data, 1981, N = 349)

labor. More than three-quarters of on-farm labor in a sample of seven households in 1981–82 came from within the household. The remaining labor was comprised of daily wage laborers (8%) and reciprocal labor groups (15%). Households possessing a large number of active workers are more capable of completing tasks in a timely manner than labor-poor households. They can also participate more easily in reciprocal labor groups which require that a household actively work in the fields of other group members. The loss of an active worker from labor-poor households is more acutely felt than it is for labor-rich households.

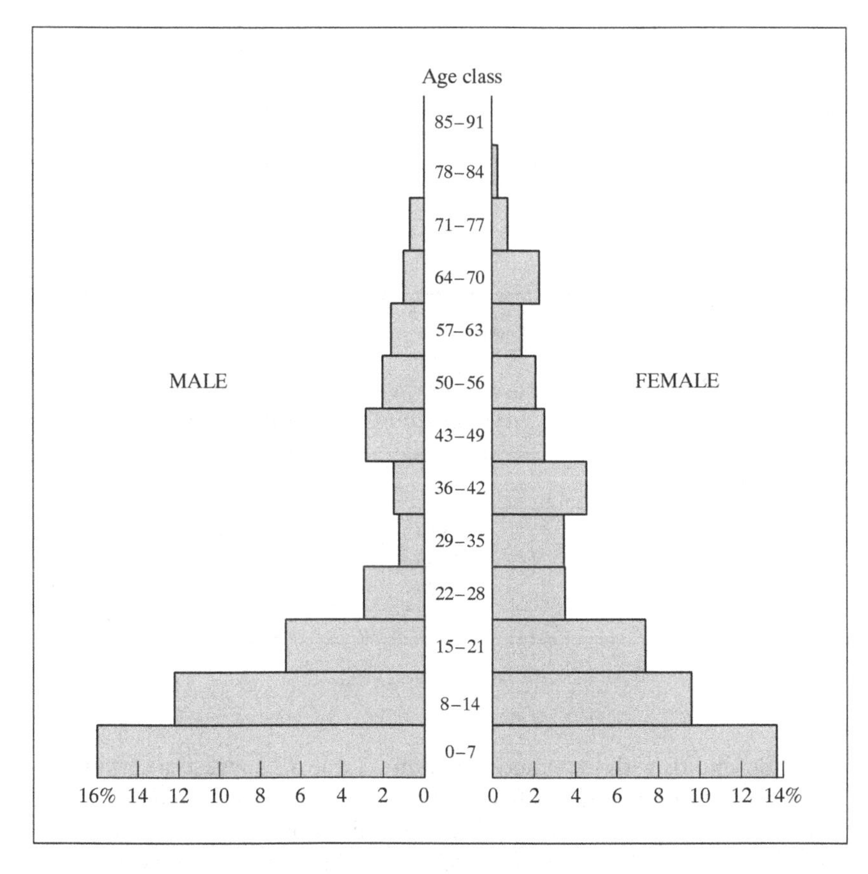

Fig. 5.6 Distribution of population by sex and age class, Katiali, 1992 (*Source:* Field data, 1992, N = 498)

The capacity of relatively prosperous households to hire in labor and to rent tractor and ox-plow services in 1981–82 is shown in Table 5.2. This was particularly true for Jula households who spent, on average, more than five times the amount of money than Senufo households.

In 1994/95 peasant farmers continued to rely upon non-household labor for farm work. Middle- and upper-income households employed, on average, two and one half to three and one half times more laborers than lower-income households. Reciprocal labor groups (Senufo: *n'golon*) continued to form the most important source of off-farm labor. The larger number of active workers in upper- and middle-income households (Table 5.1) means that they can engage in this type of labor exchange more easily

Table 5.1 *Average number of household residents and actives by income group: Katiali, 1994/95*

Income Group	Residents	Actives
I	25	13
II	13	7
III	7	3
Average	14	8

Source: Field data, N = 38 households

Table 5.2 *Average household expenditures for hiring wage labor, tractors, and animal traction (in FCFA): Katiali, 1981–82*

Income Group	Senufo	Jula	Village
I	22,835	96,600	68,490
II	4,105	28,550	15,685
III	9,820	22,710	15,680
Average	8,865	44,920	26,895

Source: Field data: N = 28 individuals

than households with fewer active workers. Table 5.3 suggests a certain leveling of household differentiation between upper- and middle-income households, which is also noticeable in the adoption of ox-plows (see Table 5.6) and the use of herbicides (see Table 5.8). Off-farm labor was employed more in cotton cultivation (55%) than in food crop fields (45%). Almost two-thirds (61%) were used during the harvest labor bottleneck period; 25% were mobilized during the weeding bottleneck.

Labor bottleneck periods

A labor bottleneck can be defined as a period in the agricultural calendar when household members work "overtime" to complete a task (e.g. weeding maize) or a sequence of tasks (e.g. ridging and planting). The stakes are often high if a household fails to complete a given task on time. For example, cotton must be planted by 30 June or else CIDT extension agents and producer cooperatives will not give pesticides to growers. Yet the "food first" strategy of Senufo farmers means that they give priority to preparing and planting food crop fields before sowing cotton. Consequently, a

Table 5.3 *Average number of off-farm laborers employed by income group: Katiali, 1994*

Income Group	N'golon	Wage	Total Average
I	180	33	214
II	133	33	167
III	44	15	59
Average	124	29	153

Source: Field data; N = 38 households

series of activities must be completed by the end of June to ensure that both the food crop and cotton yields are not compromised.

Peak labor periods occur throughout the rainy season and well into the dry season. They can be distinguished in both quantitative and qualitative terms.[33] Senufo farmers identify labor bottlenecks in qualitative terms with the expression *falou nehin* – "there is too much work." Senufo men distinguish three such periods. The first encompasses the early rainy season period when farmers are pressed for time to plow, ridge, and sow their fields (the planting bottleneck). The second peak period occurs when maize is about a meter high and farmers are occupied with thinning, weeding, and re-ridging their rice, maize, and cotton fields (the weeding bottleneck). The third bottleneck arises towards the end of the rainy season and during the early dry season, when cotton and swamp rice are harvested before their quality diminishes (the harvest bottleneck). In addition to these three peak labor periods, Senufo women identify a fourth bottleneck period associated with the transplanting of swamp rice in August and September (the transplanting bottleneck). As suggested above in the discussion on rainfall variability, the severity of bottleneck periods is reduced or heightened by rainfall patterns. If the onset of rains is late, field preparation and sowing are compressed within a shorter time frame. When rains end early, swamp rice fields must be harvested earlier than normal to prevent too steep a fall in the water content of rice. If rice is too dry when harvested, it will pulverize under the impact of a pestle in a mortar.

An outsider recognizes a bottleneck period by the relative absence of people in the village. During peak work periods, it is common for people to leave for their fields between six and seven in the morning and not return to the village until eight o'clock at night. Work is constant until the midday meal taken around 3:00 p.m. If a household's fields are distant from the village, the head will construct a field camp where the family spends the

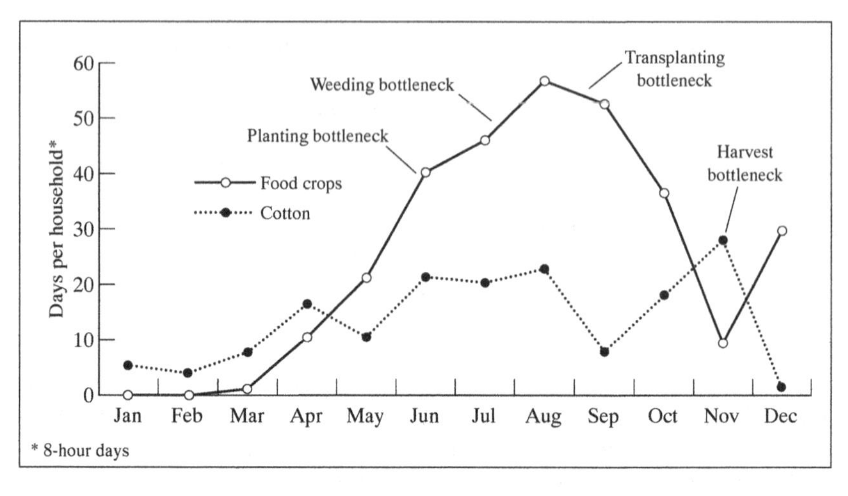

Fig. 5.7 Labor bottleneck periods, Katiali, 1981–82 (*Source:* Field data)

night. This way, time spent walking to fields is economized and family members can keep a closer watch on their crops to minimize losses from birds, monkeys, and stray cattle. For those returning to the village each night, an evening meal is consumed between 9:00 and 10:00 p.m., followed by deep sleep. At the end of these long work days, the Senufo state: *min bara n'solo*, or "we have worked too hard." This qualitative measure of a bottleneck period is supported by quantitative data summarized in Figure 5.7. The graph shows the four bottleneck periods identified by informants in relation to the number of days per month devoted by each household to food crops and cotton. How have households of different economic standing sought to smooth these peak periods in the agricultural calendar?

Farmer adjustments to labor bottlenecks

One can group a variety of adjustments into one of three realms: labor mobilization, agricultural practices, and cultural change.[34] Before proceeding to discuss these coping strategies, it is important to stress two points. The first is that labor bottlenecks are nothing new to Senufo and Jula farmers. Colonial officials observed as early as 1912 that it would be difficult to expand cotton cultivation among the Senufo without improving labor productivity. Otherwise, there simply was not enough time in the agricultural calendar if farmers gave priority to food security. In the early 1960s, SEDES similarly reported that "the work calendar is severely constrained in the middle of the rainy season" and that one could not expect to increase output without a corresponding increase in the productivity of labor.[35] The SEDES farming systems data for 1962 discussed in the following section provide a benchmark against which one can measure the nature of household adjustments to the intensification of cotton.

The second point is that differential access to the factors of production, particularly labor and capital, largely determine a household's ability to cope successfully with peak labor periods. As the following discussion shows, upper- and middle-income households are in a better position to deal with labor bottlenecks than lower-income households. Yet such differences appear to have narrowed over time, as even the poorest households gained access to labor-saving technology.

Prior to the intensification of cotton, the farming system of the Katiali area was similar to that described for the "millet zone" in the SEDES study for the year 1962. In contrast to the 1980s, food crops dominated the farming system. Table 5.4 lists the top five crops and associations for the zone in 1962 and for Katiali in the 1980s and early 1990s. The dominant position of maize and millet in the 1962 farming system reflects the largely food crop orientation of savanna farmers in this period. Cotton's leading position in the 1980s–90s illustrates the dramatic reorientation of production for the market. Three patterns stand out in this table.

The first and most striking change is the rise in cotton and rain-fed rice and the decline in millet. The crop that most distinguished the northern Korhogo region in the early 1960s was hard to find twenty years later. Millet (*Pennisetum typhoides*) is native to Africa and likely originated in the West African savanna.[36] The small, pencil or candle-shaped variety (*Pennisetum americanum*) figures prominently in Senufo cultural history as the most important grain crop grown in collective fields.[37] Coulibaly calls it the "noble cereal" because of its customary control by elders in the past.[38] The grain was stored in *segnon* granaries controlled by the *katiolofolo* who controlled its distribution. It was served in beverage form to the members of off-farm labor groups working in the *segnon* as well as to visitors upon entering the village. It was also consumed by household members on *segnon* work days in the form of a "paste" seasoned by a piquant sauce. Senufo women fermented the grain for making beer (*soum* or *chapalo*). Its cultivation was forbidden in the personal fields (*kagon*) of individual household members. This prohibition served to reinforce the social hierarchy in which the gerontocracy controlled, through various forms of power relations, the social means of production and reproduction.

Millet also possessed a "sacred" quality. At one moment in the highly ritualized Senufo funeral, the lineage head of the deceased offers millet "in the guise of a meal" to the grave diggers.[39] So important was millet that its name appears in a number of Senufo calendars. Among the Kasambélé

Table 5.4 *Top five crops and associations: Millet Zone (1962) and Katiali (1981, 1988, 1991, 1994)*

Rank/Year	1962	1981	1988	1991	1994
1	Maize-millet (48%)	Cotton (44%)	Cotton (47%)	Cotton (40%)	Cotton (34%)
2	Maize-sorghum (8%)	Swamp rice (16%)	Maize (19%)	Maize (20%)	Maize (24%)
3	Swamp rice (7%)	Maize (14%)	Rice-maize (8%)	Rice-maize (10%)	Rice-maize (15%)
4	Maize-cotton (4%)	Maize-millet (5%)	Swamp rice (6%)	Swamp rice (7%)	Swamp rice (6%)
5	Maize-millet-peanuts (4%)	Peanuts (4%)	Peanuts (4%)	Peanuts (6%)	Upland rice (4%)

Source: SEDES, *Rapport Agricole*, p. 181; field data.

Senufo, the name for July is *Koromonon*, or "the time to sow millet"; among the Kiembara, the name for January is *Sounirou* ("the millet is ripe") or *Soukayègue* ("the month to harvest millet"). [40]

The dramatic decline in millet cultivation represents a significant change in the culture of production. That is, the cultural forms that sanctioned a particular set of power relations such as millet cultivation in *segnon* fields gradually changed as access to the means of production and reproduction shifted from collective to more individual lines. Millet cultivation lost both its material and symbolic importance as the authority of elders declined, and juniors invested in crops like cotton and maize to ensure their autonomy. This observation begs the question of why maize, "the commoner's crop," has supplanted millet as the most important food crop in the Korhogo region.[41] The answer to this question takes us to one of the central dynamics of the process of agricultural intensification since the early 1960s – the resolution of labor bottlenecks in a farming system. Millet is cultivated as a second-cycle crop that is planted, thinned, and transplanted in July–August and harvested in December–January. That is, its growing cycle falls squarely within the weeding and harvest bottleneck periods in the agricultural calendar. Maize is an appealing alternative food crop for three reasons. First, it is planted as a first-cycle crop in May–June and can be harvested during August, when food stores are typically low. Thus farmers are better able to endure the "hungry season" and simultaneously reduce the harvest bottleneck. Second, from an agro-ecological perspective, maize yields are higher than those for millet in the humid Sudanian savanna characteristic of the Korhogo region. This performance edge is linked to the better match between maize's growth requirements and rainfall patterns in the region.[42] Third, maize husks better protect the grain from water, bird, and insect damage than millet, whose grains are more open to attack. Crop damage caused by transhumant cattle herds is an additional factor cited by farmers in explaining the decline of millet. Millet is left to dry on stalks in farmers' fields during December and January, which is when Fulbe herds move frequently throughout the region in search of water and pasture. The bulk of crop damage occurs during these early dry season months when range quality is rapidly declining.[43]

A second and equally striking pattern that stands out in Table 5.4 is the increase in monocropping and corresponding decline in intercropping. This is evident not only in the case of cotton but also for maize and rice. In 1962, 82% of the area was planted in mixed crops, 18% in sole crops.[44] By 1994 the ratio was just the opposite. The intercropped area had dropped to just 18% of the total cultivated area while monocrops covered 82%. The decline in intercropping is largely due to the introduction of high-yielding varieties (HYV) of cotton and food crops (maize, rice) by agricultural extension agents. These agents advise farmers to plant HYVs as sole crops based on performance trials on experiment stations. Farmers offer additional explanations for the decline in intercropping, most notably the lack of time to plant intercrops (see below).

A third pattern that appears, particularly for the period 1981–94, is the increasing area in rain-fed rice and the decline in swamp rice. New, early maturing varieties of rain-fed rice (e.g. Iguapé) are cultivated as both intercrops (rice–maize) and sole crops. Their expansion is linked to the widespread adoption of ox-plows and, most importantly, to their better fit into the agricultural calendar. Rain-fed rice is harvested in October well before the cotton harvest. Swamp rice, on the other hand, is harvested at the same time as cotton.

These three major changes in cropping patterns represent just a few of the many adjustments cotton growers have made to cope with peak labor periods in the agricultural calendar. How have households of different economic standing dealt with the chronic problem of labor bottlenecks, and how have these coping strategies contributed to the intensification of cotton in northern Côte d'Ivoire? The most common strategies involve the mobilization of non-household labor, the modification of agricultural practices, and the adjustment of ritual calendars.

Labor mobilization

One way to complete a series of competing agricultural tasks is to mobilize additional labor power from both within and outside the household. Past forms of labor mobilization that were organized around the lineage-level production unit (*katiolo*) and the institution of *poro* have increasingly given way to conjugal household arrangements. Today, the most common means of mobilizing labor is through participation in reciprocal labor groups (*n'golon*), hiring daily wage laborers (*tiowa*), and attracting suitor work groups (*léhéré*). To prevent the loss of labor from households through migration and divorce, household heads also offer incentives to their sons and wives to remain attached to their production units as active workers.

Before the breakup of the *katiolo* production units in the early 1940s, labor was mobilized at the lineage (*narigba*) level. The large collective fields (*segnon* or *segbo*) cultivated by lineage members and their attached dependants benefited from a large pool of workers. A major incentive for young men to work in the segbo was the promise of a bride to be given to him by the head of the *katiolo* (*katiolofolo*) called *segnontio,* or "a bride of the *segnon*". With the break up of the segnon, a man seeking a wife was increasingly forced to win the favor of individual household heads. One common strategy was to work in the fields of prospective in-laws in the hope of being rewarded with "a wife for the good he has done," or *katiene-tio*. A suitor typically recruits ten to twenty friends to work one day a year in their future in-laws' fields. The field owner provides a copious meal to the members of these *léhéré* work parties. Of course, the suitor must reciprocate and work one day in the fields of each of his friends' future in-laws.

N'golon work parties are similar to the *léhéré* groups in that they are based on the principle of reciprocity. Typically, a household head who is anxious to complete a specific agricultural task will recruit non-household workers to work in his field on a specified day. A generous midday meal is prepared by the inviting household to feed this group, which will work from nine to ten in the morning to early evening. The organizer of this work group is obliged to pay back each member of the *n'golon* work group by working in their fields. *N'golon* labor constitutes an important source of non-household production. In 1981–82, off-farm labor accounted for one-quarter of total agricultural labor time for the seven households participating in the farm management study. Two-thirds of this labor was recruited as *n'golon* labor. The remaining third was composed of daily wage laborers (*tiowa*), mainly women and children, who commonly worked alongside *n'golon* work groups for the equivalent of one dollar a day. Household expenditures on *tiowa* labor quadrupled in real terms over the twenty-year period between 1962–82. In general, off-farm labor was most often recruited to prepare swamp rice fields during the weeding bottleneck and to harvest both food crops and cotton during the harvest bottleneck.

Women also organize *n'golon* labor parties to work in their individual fields during bottleneck periods. If they have the means, they also employ wage laborers, and they sometimes pay their spouses to plow their fields. Some women also organize themselves into *tiowa* work groups which function like a rotating savings and credit group. A group hires itself out for a certain fee to a household seeking off-farm labor, but just one member of the work group takes home the entire payment. On the next work day, a different member of the group receives the payment. Each member of the group is obliged to work on contracted dates until everyone has received the lump sum. The appeal of *tiowa* work parties is twofold: its members have access to a relatively large sum of money at once, and labor-seeking households save time by contacting the head of such a group rather than recruiting individual daily wage workers.

In addition to recruiting non-household labor, household heads seek to keep family members attached to the household as active workers. A major concern of household heads is that their sons or nephews will leave to work outside the village. As an incentive to keep potentially mobile members attached to their production unit, household heads commonly allow their sons and nephews to cultivate an individual cotton field. The struggle to

Plate 17 A mixed *n'golon* and wage labor group harvesting cotton, Katiali, January 1982

maintain control over this important segment of the work force reached crisis proportions in the early 1970s, when household heads supported a revolt of *poro* initiates (*tyolobélé*) against the labor exactions of the village chief, Zanapé Silué. The *tyolobélé* revolt is an example of how Senufo farmers have reshaped cultural traditions to cope with the problem of labor bottlenecks.

The tyolobélé *revolt*

In the 1960s, the *tyolobélé* worked two days a week in the fields of the village chief, N'crin'golo Silué. According to N'golofaga Silué, a poro initiate at the time, if N'crin'golo was satisfied with the *tyolobélé*'s work on the first day, he didn't ask them to work the second day.[45] When Zanapé Silué assumed the chieftaincy in 1970, he continued to receive the free labor of the *tyolobélé* in his personal fields. For the three years 1970, 1971 and 1972, they worked two days a week during the rainy season in his fields. In July of 1972, Zanapé broke this tradition by requesting that the *tyolobélé* work more than two days per week. He called on them whenever there was too much work to do in his fields. Zanapé typically summoned the head of the *tyolobélé* to his compound the night before he wanted the young men to work in his fields. This intermediary then informed individual *tyolo* about the next day's activities. The *tyolobélé* and their elders became increasingly critical of Zanapé's exactions. N'golofaga Silué explained the reasons for their discontent:

Working all the time in Zanapé's fields didn't leave us with enough time for our elders' fields. Also, if we wanted to travel we had to go tell the chief that we couldn't work that day; if we didn't go see him, he would fine us. One day we decided that we were not going to work in Zanapé's fields more than two days a week. We decided this when Setionwanzié (the head of the *tyolobélé* cohort and the chief's intermediary) was not around because we thought he would disagree with our actions. So we held a secret meeting when it was time to go work in the chief's swamp rice field and decided not to go. Setionwanzié went to the chief's field that next day and he was the only one there until noontime. He stayed until the evening, working by himself. We told the *tyolobélé* group that graduated before us and they agreed with our actions, which encouraged us. . . .[46]

Four days later Wamana (the representative of the dissident *tyolobélé*) traveled to Korhogo to inform the "canton chief" (Bema Coulibaly) about what was happening.

Zanapé was not happy and we didn't want him to send someone first. Bema summoned Megnergué (the head of the *tyolobélé* group that preceded the current group), Setionwanzié, Zanapé, and Wamana to a meeting. Bema did not agree with Zanapé that we should work more than two days a week. He said that the *tyolobélé* should only work two days otherwise the *poro* would be too hard and others would not want to be initiated. We agreed to work two days a week, but when we returned Zanapé was discouraged and said that he didn't care if we worked in his fields or not. It ended that way. Setionwanzié no longer came by the night before to tell us that we had to work in the chief's fields the next day.[47]

Zanapé was not a popular chief. But this story illustrates more than antipathy towards autocratic rule. For the purposes of this study, it demonstrates how control over the labor power of young men became a point of contention between the village chief and the elders of the *tyolobélé* who supported their struggle. Underlying the *tyolobélé* revolt was the view shared by initiates and their elders that an inordinate amount of time was being spent in the chief's fields to the detriment of household fields. As N'golofaga summarily put it: "we didn't have enough time." The end of the institution of *tyolobélé* labor in the village chief's fields must be seen as a victory for the household heads who wanted to channel the labor of young active male workers onto household fields. This dramatic example of a cultural change affecting the capacity of households to cope with labor bottlenecks is complemented by other, more subtle strategies that serve to free labor power when it is needed most.

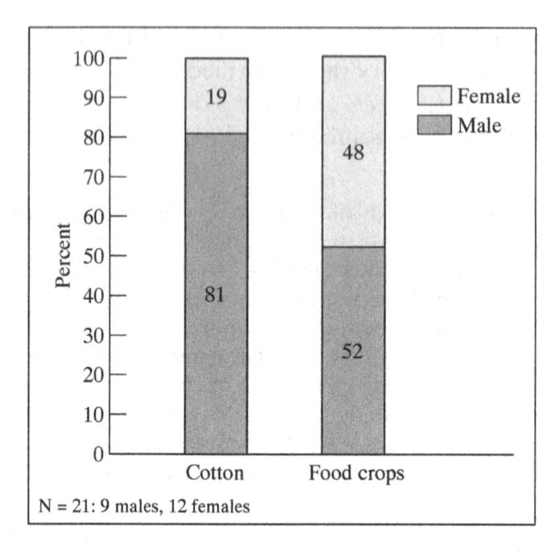

Fig. 5.8 Household on-farm labor devoted to food crops and cotton by sex, Katiali, 1981 (*Source:* Field data)

Women's agricultural work

A more subtle but equally important change in labor allocation towards household production is the increase in women's labor time to household fields. The results of a labor allocation study undertaken with nine men and twelve women in 1981–82 suggest that women were contributing substantially more time to agriculture than they had been twenty years earlier. The most dramatic difference appears in the total number of days devoted to agricultural work. In comparison to 1962, when women of the Korhogo region spent 90 days per year engaged in agricultural work, women in the Katiali region devoted 145 days to agriculture – an increase of nearly 60%. Men's agricultural labor time increased by 29%. This increase in women's agricultural activities is in addition to the already burdensome time they devote to child care and household work (146 days versus 5 days for men). This observation on the general feminization of agriculture in the case of Katiali is supported by the findings of other researchers in the region.[48]

Figure 5.8 shows the distribution of women's and men's labor time to food crop and cash crop production. Women provided 42% of the total household labor time. Senufo women devote more time to food crop production than Senufo men. Dry season tobacco cultivation, an ethnoeconomic specialization of the Jula in the region, was dominated by Jula women. Men of both ethnic groups devoted the bulk (81%) of household

Table 5.5 *Senufo days of the week*

Noupka	Tori	Kali	Tiefonon	Koundiali	Kong

labor time to cotton cultivation. Although women provided just 19% of total household labor inputs to cotton, they comprise an important part (one-third) of the off-farm work force which is heavily involved in cotton cultivation.

The increase in labor time devoted to agriculture by both men and women is partly due to the expansion in total cultivated area since the early 1960s. The SEDES study of the Korhogo region for 1962 reported an average cultivated area of 3.27 ha for households using manual farming techniques. Eponou noted that the cultivated area in M'Bengué in the early 1980s had increased to 5.2 ha for manual households, 8.8 ha for households using ox-drawn plows, and 35.3 ha for tractor-owning households.[49] Since less than 1% of the households in the Korhogo region farm with tractors, one can conclude that the increase in women's work has been associated with the expansion of cotton among manual and animal-traction households.

The number of days women are allowed to work in individual (versus household) fields varies between ethnic groups and sometimes among households within the same group. Senufo women have the right to work in their personal fields two days (*koundièle* and *kong*) in the Senufo six-day week. Most senior married women also enjoy a rest day, usually Friday. On the remaining days, they work in the household's fields. On days that they work in their individual fields, Senufo women must provide their own food. The household head is responsible for feeding household members only on days they work in household fields. Jula women, conversely, do not have any fixed work days. They work in the household's fields when asked by the household head (usually their husband) but are otherwise free to work in their own fields, if they farm at all.

Women of both groups do not have free access to the labor power of household members. Even if mothers want their own children to work in their individual fields, they must obtain their husbands' permission. The only exception is for nursing mothers, who are normally allowed to take one of their daughters with them to their fields to help with child care. This restriction on access to household labor leads to the frequent employment of off-farm labor on women's fields.

In 1981–82, women usually cultivated an upland peanut field and a lowland swamp rice field. Only a few women cultivated cotton, sometimes with their co-wives, and with a heavy reliance on off-farm labor. The land area cultivated by women amounted to 14% of the total cultivated area.

This percentage declined when the cotton area per active worker increased within the household.

In summary, by the early 1980s, the combined effects of labor migration, the intensification of cotton, and the limited development of labor-saving technology led to an absolute increase in the agricultural labor time for both men and women. This increased work burden fell most heavily upon women. The 1981–82 data suggest that the increase in women's agricultural labor time entailed a shift from working in their personal fields to working in household fields and in non-household fields. This contrasts sharply with the early 1960s, when women enjoyed considerably more autonomy and dominated market transactions.[50]

Manipulation of cultural calendars

Senufo farmers manage to cope with peak labor periods by their flexible interpretation of culturally-prescribed calendars such as the observance of days of rest, funeral ceremonies, and women's work days. Two types of rest days are observed by Senufo farmers: personal rest days (*tiandin*) and field rest days (*segui tiandin*). Personal rest days are designated by a diviner (*sando*), who informs a farmer that he or she should not work on a certain day. Individuals commonly consult diviners to determine the cause of specific problems (e.g. illness, miscarriages, poor harvests) and what they must do to obtain a desired result. During a divination session, a *sandogo* communicates with bush spirits (*tugubélé* or *madebélé*), who may indicate that the client must observe a personal rest day (e.g. *Tori*) to improve their situation. Anita Glaze describes the links between bush spirits, divination, and ritual observances in her book *Life and Death in a Senufo Village*:

> The bush spirits are thought to be the chief source of the *Sando*'s power and the chief cause of her clients' problems. *Madebele* are animate spirits who inhabit the bush, fields, and streams surrounding the villages. They demand gifts, worship, and visual commemoration in order to remain happy. If one unwittingly disturbs them or breaks a promise made to them, they can become angry enough to kill. Cultivating a field where there are "ancient settlements" of *madebele* can be an especially dangerous risk for farmers, who usually consult *Sando* diviners as a preventive measure. The diviners advise the farmers to sacrifice a chicken and "talk with the *madebele*" in their fields.[51]

A farmer will also consult a diviner before opening up a new field from the bush in order to avoid any unforeseen problems. It is at this time that the *Sando* will inform a client about any obligatory field rest days or *segui tiandin*. These refer to *places* where farmers are forbidden to work on a specific day. A farmer might also learn about field rest days from a lineage head or the village earth priest (*tarfolo*), who regularly makes sacrifices to the spirits dwelling in the area where the new field will be located. Such observances are meant to appease the powerful bush spirits that reside in that area. One courts misfortune by ignoring such prohibitions or by failing to perform a sacrifice.

Taken together, these time- and place-specific prohibitions against work pose potential obstacles to farmers seeking to reduce labor bottlenecks. However, the Senufo's flexible interpretation of these restrictions means that a day's rest is not a day's labor lost. For example, although individuals are prevented from working in their fields on personal rest days, this does not mean that other family members are also prohibited from working. The prohibition is interpreted as a personal restriction that does not apply to other family and non-family members. One can even hire wage-laborers and be present in the field to offer food and water on personal rest days.

A similar flexibility is apparent in the interpretation of restrictions associated with field rest days. Farmers who adhere to traditional religious practices will not work in a field during its *segui tiandin*. However, it is common for farmers to have fields in a variety of locations that have different rest days. For example, the farming area known as Nafoulagba has *kali* for its rest day, while Friday is the *segui tiandin* for the area known as Planigué. Therefore on Friday the farmer can work in his Nafoulagba fields and on *kali* work in fields located at Planigué. The more dramatic but not uncommon practice of religious conversion (typically to Islam) frees the convert from having to observe both personal and field rest days. This is apparent among the Muslim Jula, who disregard an area's *segui tiandin* in their weekly calendar.

The postponement of funerals to the dry season is another example of the Senufo's flexible interpretation of ritual calendars. Funerals are major ceremonial events that last a minimum of three days. To ensure that a proper ritual funeral takes place requires considerable time and financial resources. Both are usually in short supply during the rainy season, when death is no stranger. Typically, rainy season deaths involve a formal burial, but the ritual funeral is held during the dry season, when time and resources permit a proper funeral.[52]

A similar flexibility is shown by women farmers when they choose to work in household fields on days that are designated as personal field days. Like other farmers, women face considerable time constraints in cultivating their fields. Unlike male household heads, they do not command the resources that might enable them to surmount labor bottlenecks during the agricultural calendar. Yet during peak labor periods it is not uncommon to find women working in household fields on *koundiali* and *kong*. This is particularly true for households in which the husband and wife have a good relationship. Women who choose to spend their individual field work days

Table 5.6 *Percentage of households owning oxen*

Income group	1981	1985	1988	1991	1994
I	63	86	100	100	100
II	37	63	89	95	100
III	18	36	75	82	75
All groups	37	61	87	92	95

Source: Field data; N = 38 households

in household fields invariably receive some compensation. In addition to eating a meal prepared from household (versus personal) food stores, women may be compensated by having their personal fields plowed by the household ox-team, greater access to their children's labor, or other assistance with field work.

All these examples show that peasant farmers are constantly adjusting their schedules and bending the framework and meaning of culturally prescribed calendars in order to cope with labor bottlenecks. The flexibility displayed in the cultural realm is matched by an equally impressive set of innovations in farming techniques. In the end, it is the combination of and dynamic interplay between all of these initiatives aimed at breaking bottlenecks that have permitted the insertion and ultimate expansion of cotton in peasant farming systems at the village level.

Agricultural practices

Peasant farmers smooth labor profiles by adopting labor-saving techniques, by switching to early maturing and less labor intensive crops, and by neglecting certain tasks because of insufficient time. The two most common techniques for increasing labor productivity are the use of ox-plows and herbicides. Ox-plows in particular have expanded at an impressive rate over the 1980s. Table 5.6 shows the growing percentage of households owning ox-plows in Katiali over the 1980s. This impressive increase in animal traction has helped cotton growers to manage the integration of cotton into their farming system. The use of animal traction for plowing and ridging has allowed peasant farmers to break up the planting bottleneck. Farmers estimate that it takes just five to six days to plow and ridge a one-hectare field with oxen, in contrast to 30 days of work if done manually.[53] When one includes the time saved in the even more laborious task of plowing swamp rice fields,[54] it is believed that ox-plows lead to a savings of 38 days per

Plate 18 Donisongui Silué and son, Ténéna, plowing a field with oxen, Katiali, June 1986

Plate 19 Donisongui Silué preparing a portable (ultra-low volume) herbicide sprayer, Katiali, June 1986

active worker in field preparation. However, to use ox-plows efficiently, fields must be destumped – a new and very time consuming task that takes place during the dry season. Peltre-Wurtz and Steck estimated that approximately 30 days of work per agricultural active are taken up in the task of destumping, which involves mechanically removing the roots of trees that have been cut down when preparing a field for cultivation.[55] Thus, it appears that 30 days of work have been transferred from the rainy to the dry season as a result of animal traction, and that eight days of (men's) labor have been economized. The point is that the adoption of ox-plows considerably lightens field preparation time during the planting bottleneck.

These findings support the view that labor productivity increases in the transition from hand hoe to plow cultivation, at least for some activities like land preparation.[56] However, there are good reasons to believe that the total labor requirements per hectare do not decline with the adoption of the plow. In addition to Peltre-Wurtz and Steck's observation on destumping, Bigot believes that the time savings in field preparation are consumed by more time allocated to weeding and harvesting.[57] This seems to be the case when weeding and harvesting continue to be performed manually. Evidence from Katiali suggests that the area under cultivation per active worker has stagnated despite the transition from the hand hoe to the plow. As we will see below, the increasing use of herbicides indicates that farmers are continually searching for ways to break up the weeding bottleneck.

Even households that did not own oxen were able to employ an oxen team in some of their fields. In 1981, 13% of the sampled households hired an ox-team and driver for plowing, weeding, and ridging.[58] In 1991, the three households that did not own oxen still managed to have at least one of their fields plowed by oxen. One widow had her swamp rice field plowed as a gift. A divorced woman also had her one-hectare cotton field plowed as a gift by an ox-team owner. However, she had to provide in-kind services to a second ox-team owner who plowed her half-hectare maize-sorghum field in exchange for fifteen days of work harvesting cotton. The third household made an arrangement with an ox-team owner who did not have a son old enough to lead his oxen nor to look after them during the dry season. In exchange for having his two and a half hectare field plowed by oxen, the field owner committed his son to work for the ox-plow owner for the entire year.

Table 5.7 shows that farmers employed a variety of techniques to cultivate their fields. When a tractor was available for hire, upper income households most commonly exercised this option. In 1981 tractors were used on 10% of the area cultivated at a cost of 20,000–30,000 FCFA ($86–120) per hectare, depending on the tasks performed. Two-thirds of the households

Table 5.7 *Cultivated area by technique: Katiali, 1981, 1988, 1991, 1994*

Tech./Year	1981	1988	1991	1994
1. Manual	41%	4%	6%	2%
2. Ox-plows	46%	96%	86%	97%
3. Tractor	6%	—	1%	1%
4. Man./Ox-plow	1%	<1%	1%	<1%
5. Man./Tractor	<1%	—	—	—
6. Ox-plow/Tractor	4%	—	7%	—
7. Man./Ox-plow/Tract.	2%	—	—	—

Source: Field data; N = 38 households

Table 5.8 *Percentage of households using herbicides by income group: Katiali, 1986, 1988, 1991, 1994*

Income group	1986	1988	1991	1994
I	63%	100%	100%	100%
II	32%	95%	95%	82%
III	36%	73%	73%	50%
Average	40%	89%	89%	79%

Source: Field data; N = 38 households

hiring tractors belonged to the upper-income group with the remaining third from middle-income households. In 1991, tractors were hired to plow about 8% of the cultivated area. Again, upper-income group households accounted for most (three-quarters) of tractor use and middle-income households for the remainder (one-quarter). The decline in area worked by tractors after 1991 is linked to the worsening economic conditions discussed in detail in chapter 6.

Herbicides were increasingly used as a labor-saving technique to break up the weeding bottleneck occurring in the months of July and August. They were applied to plowed and sown fields to suppress weed growth for up to sixty days. After that, farmers manually weeded their fields.[59] Table 5.8 shows that households across income groups were applying herbicides on their fields. The table includes any household that used a herbicide on at least one field during that year. A growing proportion of households were using herbicides throughout the 1980s and early 1990s. The number began to drop, however, as the cotton economy deteriorated as suggested by the

Table 5.9 *Average area sprayed with herbicides by income group: Katiali, 1994*

Income group	Average number of hectares sprayed
I	3.28
II	2.56
III	0.90
Average	2.37

Source: Field data; N = 38 households

data for 1994. Middle-income, and especially low-income, households were particularly affected by the adverse conditions.

When one measures the actual area sprayed with herbicides, the variation among households stands out even more clearly. Table 5.9 shows that upper-income households, on average, sprayed an area that was more than three-and-one-half times larger than that of low-income households. Middle-income farmers used herbicides on an area almost three times larger than that of low-income farmers. In 1994 herbicides were used on 31% of the total cultivated area. Of the area treated, 47% was in cotton and 53% in food crops. The top three crops receiving herbicide applications were cotton (47%), rice-maize (27%), and maize (18%). These three crops or crop combinations accounted for 92% of the total area treated with herbicides. Farmers can obtain herbicides on credit from village cooperatives. They repay their loans at the time of the cotton market when the cooperative simply deducts what it is owed from the grower's earnings – assuming the farmer sells sufficient quantities of cotton.[60]

Peasant farmers also draw upon a repertoire of indigenous agricultural practices to reduce labor demands when they are most intense. For example, the weeding bottleneck can be lightened by more thorough ridging of fields. Weeds are killed or their growth suppressed when they are covered with soil heaped up by plowshares. Intercropping also effectively suppresses weed growth when the dense plant cover and root systems compete with weeds for sunlight, water, and soil nutrients. The decline in intercropping associated with the expansion of higher-yielding varieties such as cotton, maize, and rice means that this indigenous practice of weed control is declining in importance. A third technique for reducing labor inputs during the weeding bottleneck is judicious crop successions. For example, a field that has been cultivated for five years usually has more weeds than a recently cultivated field. Since a crop like sorghum is more tolerant of weeds than other crops like rice, farmers will plant crops in older parcels that are most likely to do well even if the field is not regularly weeded.

Contemporary farming practices tend to exacerbate the weeding bottleneck by stimulating the growth of weeds. Monocropping accelerates weed growth by reducing competition between weeds and secondary crops.[61] Moreover, weeds are more problematic in fields that have been fertilized in previous years. Food crop yields may be higher as a result of residual fertilizer in soils, but they also suffer from a proliferation of weeds linked to soil acidification. Sement shows that the soil acidification problem is linked to the interactive effects of fertilizer use and reduced organic material.[62] Finally, Peltre-Wurtz and Steck show that the expansion of ox-plows has led to a shortening of fallow periods because of the relative ease of destumping young versus old fallow.[63] Weeds tend to be a greater problem in fields that have had a shorter fallow period. The growth in the use of herbicides is thus linked to these environmental changes which are associated with the intensification of cotton.

Conclusion

The period 1964–84 presented a pattern of intensification driven by the interplay of directed innovations, socio-cultural change, and locally induced innovations linked to peasant efforts to reduce labor bottlenecks in a labor-constrained farming system. Directed innovations took the form of attractive administered prices in both input and output markets, producer subsidies (fertilizers, pesticides, seeds), and new seed varieties. In contrast to the coercive phase of directed innovations characteristic of the 1960s, this seduced innovation phase lured greater numbers of peasants to cotton growing. The successful adoption of the HYV package was contingent upon farmers' flexibility in modifying their socio-cultural and agricultural practices. Specifically, the adoption of new technologies (ox-plows, herbicides), shifts in cropping patterns (increase in monocropping, decline in millet), flexible interpretations of culturally prescribed rest days and funeral periods, and new forms of labor mobilization involving both contestation (the *tyolobélé* revolt) and negotiation (increase in women's work) have been central to the process of agricultural intensification. My core argument that these locally induced innovations have been driven largely by farmer concerns to level peak labor periods in a new agricultural calendar supports Sara Berry's view that agricultural intensification in sub-Saharan Africa has often "occurred in response to increasing constraints on farmers' time, rather than growing scarcity of land."[64]

The Katiali case study indicates that the cotton boom fundamentally altered the structure of the rural economy and society in northern Côte d'Ivoire in at least three ways. First, the heightened interest in cotton growing produced tensions among and between generations over the control of labor, as witnessed by the *tyolobélé* revolt. The end of *tyolobélé* labor service in the fields of the village chief represented a further erosion of village and lineage-level labor control institutions that had been in the process of dissolution since the 1930s and 1940s. Their replacement by the conjugal family as the dominant social unit of production accelerated during this period. Second, women's labor inputs into farming increased more than those of men. The time allocation study showed that the burden of smoothing labor profiles fell disproportionately on the shoulders of women. In contrast to the early 1960s, women were devoting more time to household fields controlled by men and less time to their individual fields.

Third, the process of agricultural intensification was uneven among households due to differences in access to productive resources. More prosperous households were better positioned than lower-income households to cope with the labor supply problems accompanying the cotton boom and exacerbated by high emigration rates. They were the early adopters of ox-plows and herbicides, and employed wage laborers more often than lower-income households. Non-oxen owning households had to barter their labor or their children's labor in order to have their fields plowed. Upper-income households disproportionately possessed the means to hire tractors as opposed to middle- and lower-income households. As a consequence of better access to productive resources, active workers in wealthier households cultivated a larger area than active workers in poorer households.[65] The larger number of active workers residing in upper-income households also allowed them to employ reciprocal labor groups more often than labor-deficient, low-income households. As economic conditions worsened in the second half of the 1980s and early 1990s, they were also more successful in avoiding debt.

The steady decline in producer prices in constant FCFA during this period did not deter farmers from continuing to grow cotton. In the absence of an alternative cash crop, they were drawn to the multiple subsidies, timely delivery of inputs and cotton earnings, and a credit system enabling them to transform their farming practices. The ability of increasing numbers of farmers to obtain ox-plows permitted many to increase not only their cotton area but also their area in food crops. Indeed, one of the main attractions of cotton growing to many farmers, especially those residing in low- and middle-income households, was that it was the only way they could secure access to the means of agricultural intensification (e.g. fertilizers, herbicides, ox-plows). Although these new techniques were first applied to cotton, farmers increasingly used them on food crops.[66] The dynamics of cotton cultivation took a new turn in the second half of the 1980s when macro-economic policy reforms resulted in peasants bearing an increasingly heavy and disproportionate share of the burden of structural adjustment. The following chapter examines how the pattern of agricultural intensification discussed above was to change to one of increasing extensification between 1985/86 and 1994/95 as the conditions of production and exchange worsened and farmers modified their farming practices to cope with the growing costs of cotton.

6 "To sow or not to sow": the extensification of cotton, gender politics, and rural mobilization, 1985–1995

CIDT is doing better but it is still vulnerable to the extent that it depends on a crop cultivated by a large number of planters, about 150,000 growers each year. Thus, their decision to sow or not to sow can have very serious consequences on the operation of this business.

Coulibaly Samba, General Manager, CIDT (*Fraternité Matin,* 9 July 1995)

If I could divide myself into six persons, it would be a lot easier.

Katienen'golo Silué, Katiali, 1986

Koronden sene banza! [Jula: "Cotton growing is a gift (to CIDT)!"]

Bema Koné, Katiali, 21 August 1995

A world glut in cotton production in 1985/86 marked a turning point for the Côte d'Ivoire cotton sector. China's harvest was the largest ever, while cotton production rebounded in the United States after precipitously dropping in 1983.[1] The world price for a kilogram of lint cotton dropped from 16.55 French Francs (FF) in May of 1984 to 5.50 FF in August of 1986. With CIDT's cost price running at 10.32 FF per kg, the cotton company began to accumulate massive debts.[2] Under pressure from the World Bank and bilateral donors to reduce its operating costs and to raise its profit margin, CIDT systematically cut producer prices, eliminated its pesticide subsidy, restructured its extension system, and transferred important input distribution and credit responsibilities to village marketing cooperatives.

Peasant responses to these draconian policies were threefold. The first was to modify their farming systems with an emphasis on agricultural extensification and diversification. This option was as old as forced cotton, dating back to the second decade of the century. The second was to establish tighter control over household farm labor, notably women's labor. This involved the active discouragement of women's cotton growing by male household heads. Tensions within households around this issue resulted not only in the withdrawal of women from cotton growing, but also in a decline in the number of days women were allowed to work in their personal fields. The third strategy involved an unprecedented mobilization of cotton growers, who waged strikes at village-level cotton markets. The strike activity was organized through the institutional structure of local and regional cotton production and marketing cooperatives which had been evolving since the mid-1970s. The strategies had a common economic goal of defending (male) cotton grower incomes by reducing the costs of farm inputs (fertilizers, insecticides), improving producer prices, and increasing women's labor contributions to household fields controlled by men. The emphasis on agricultural extensification contrasted with the pattern of intensification witnessed over the previous two decades. The economic policy successes of cotton grower cooperatives were made possible in part by the spaces opened up by structural adjustment (market liberalization, decentralization) and democratization (multiparty politics). The emergence of dynamic small farmer organizations represents a new chapter in the role of peasants as important actors in the agricultural policy process. One of the outcomes of the new agrarian politics has been to secure cotton growers a larger share of the agrarian pie following the dismantling of the CFDT system in Côte d'Ivoire.

The erosion of farmer incomes

Producer price reductions began in 1989 when the administered purchase price for second-grade seed cotton fell from 105 to 100 FCFA. The following year (1990) the price of first and second-grade cotton dropped from 115 to 100 FCFA and 100 to 85 FCFA respectively. Price cuts continued into the next year (1991) when CIDT fixed the first grade cotton price at 90 FCFA and the second-grade price at 80 FCFA. This third price cut followed the signing of a new agreement ("Convention-cadre") between CIDT and the Ivorian state in 1991 which required CIDT to operate like a private company and balance its own accounts without the aid of the Ivorian marketing board (Caistab). The successive price reductions followed CIDT's strategy to reduce its operating costs and to generate profits.

Farmer incomes were further eroded by CIDT extension agents' severe cotton quality grading in local markets, which led to sharp declines in the amount of first-grade cotton purchased by the cotton company. In contrast to the period 1970–87, when more than 95% of the cotton sold to CIDT was graded top quality, the proportion dropped in 1988 to 61%, and then again to 36% in 1989. In 1990 the proportion of first-grade cotton rose to 66% and then fluctuated between 88% and 93% of total cotton purchased between 1991 and 1995. Cotton growers interpreted CIDT's draconian grading policy as a little disguised cut in the guaranteed producer price.[3] These cost-cutting measures were effective. CIDT's cost price dropped from 12.58 FF/kg of lint cotton to 9 FF/kg between 1985 and 1992.

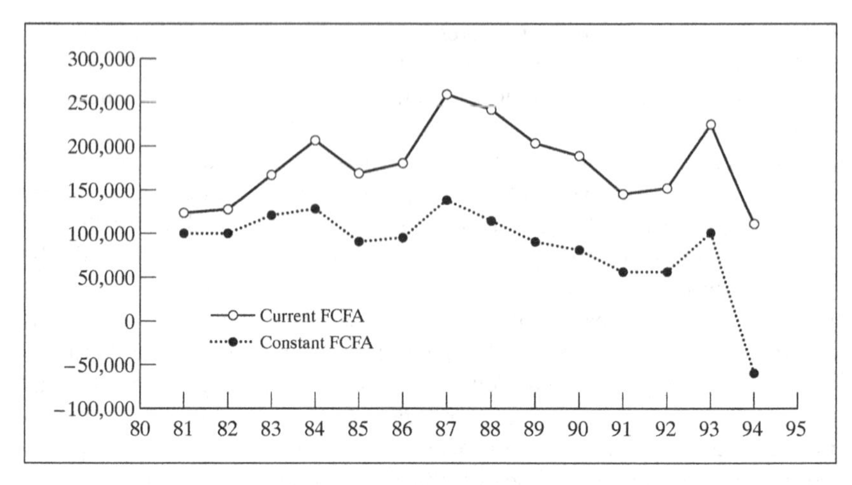

Fig. 6.1 Evolution of cotton incomes/hectare, Korhogo region, 1981–94 (*Source:* Field data and CIDT)

Further price reductions were scheduled for the 1993 campaign when the price for first-grade seed cotton was set to drop from 90 to 70 FCFA and second-grade cotton from 80 to 60 FCFA. However, in January of 1994, the Côte d'Ivoire government capitulated to International Monetary Fund and French government pressure by agreeing to devalue the West African franc (FCFA) by 50%.[4] Devaluation should have been a boon to cotton growers whose product trades in US dollars on the world market.[5] But CIDT only increased (retroactively) producer prices by 14% to 105 FCFA for first-grade cotton and from 80 to 90 FCFA for second-grade cotton. The following year (1994), the state withdrew its 100% pesticide subsidy which, in the wake of the currency devaluation, substantially raised the costs of peasant cotton production. Figure 6.1 shows that cotton incomes per hectare in both constant and current terms improved in 1993 when the pesticide subsidy was still in effect and producer prices rose. However, incomes continued their downward slide in 1994 when farmers were required for the first time to pay for pesticides. One measure of the worsening cost/price structure for cotton growers is illustrated in Fig. 6.2. The graph indicates that the number of kilograms of cotton that farmers had to sell to obtain farm inputs for one hectare of cotton significantly increased after 1994.

In addition to reducing producer prices, CIDT sought to increase its profits by improving gin yields and expanding its markets for cotton products. The cotton company unilaterally (re)introduced a glandless cotton variety (GL7) into the Korhogo region in 1988, with the goal of marketing

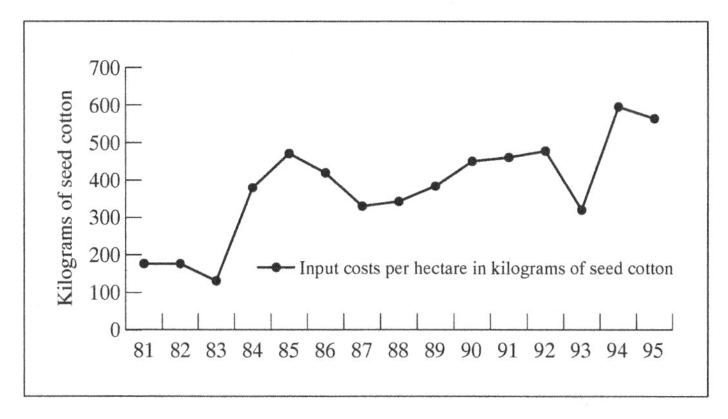

Fig. 6.2 Number of kilograms of cotton necessary to pay for recommended inputs, 1982–95 (*Source:* Field data and CIDT)

its non-toxic seed as feed grain for fowl.[6] CFDT, the foreign partner of the Ivorian parastatal CIDT, was also attracted by the higher gin yields (up to 47%) since it received a 6% commission on the value of cotton exports.[7] In experiment station field trials, GL7 attained yields of up to 3 tons per hectare in contrast to 2–2.5 tons for the ISA variety. In the tradition of directed innovation, CIDT systematically began replacing the more hardy ISA with glandless varieties in the late 1980s. The area in glandless cotton expanded rapidly in the northern region from 3% of the area in 1988 to 96% in 1993.

In contrast to CIDT's enthusiasm for glandless cotton, cotton growers were not impressed. The performance measures most important to peasants center around a variety's hardiness to variable rainfall, pests, weeds, declining soil fertility, and late planting. The question that policy makers and researchers should have been asking themselves at experiment stations was: "How will this variety perform in peasant farming systems characterized by chronic labor and capital shortages?". Given the prevalence of labor bottlenecks in peasant farming systems, farmers are more concerned with how a new variety's growing requirements complement or conflict with the growing of food crops in the agricultural calendar than with ginning yields or secondary markets for seed cotton. The more tolerant a high-yielding variety is to irregular applications of fertilizers, pesticides and/or late planting and weeding, the more attractive it is to growers. Because it proved to be intolerant to less than ideal conditions, GL7 performed poorly in peasant fields. Its longer growing season (120 days) in comparison to ISA (90 days) meant that it had to be sown earlier (by June 20 versus June 30). This requirement exacerbated rather than reduced the planting bottleneck. It

Table 6.1 *Area in glandless cotton varieties: Côte d'Ivoire, 1984–96*

Year	Area (ha)	Percentage of north*	Percentage of total CIDT area
1984	23,700	16	16
1985	995	1.4	0.7
1986	—	—	—
1987	—	—	—
1988	3,095	3	1.5
1989	9,465	9	5
1990	20,013	17	10
1991	80,568	70	42
1992	71,866	59	32
1993	128,043	96	58
1994	133,999	96	55
1995	100,660	92	49
1996	—	—	—

Source: CIDT Annual Reports; Groupe de Travail, *Coton*, 150
*includes GL7 planted in the CIDT divisions of Korhogo, Boundiali, Ferké, Bondoukou; Odienné is included for 1993–95

was also relatively more demanding than ISA in terms of its moisture requirements and the timing of pesticide applications and weeding. With the toxic gossypol removed, the plant's natural pest resistance declined, making it more vulnerable to a wide range of pests. One of the great strengths of the ISA variety was its hardiness; it was more forgiving if farmers were late in spraying or weeding their fields.[8] Because of the difficulties encountered in growing GL7 in the combined food crop/cotton farming system, yields and net cotton incomes plummeted. Realizing that it had seriously miscalculated the suitability of GL7 to peasant farming systems, CIDT decided to abandon it in favor of ISA after the 1995–96 campaign.[9]

In summary, the economic and agronomic conditions of HYV cotton production became less and less attractive to growers over this ten-year period of structural adjustment. In 1991, in the middle of the adjustment period, real producer prices were less than half their early 1970s level. According to CIDT data, cotton yields per hectare were also at an all-time low that year when the new glandless variety was planted on 42% of the total cotton area. Declining yields further eroded peasant incomes. The elimination of the pesticide subsidy, combined with the devaluation of the FCFA, dramatically increased the costs of production.

How did these increasingly adverse conditions affect the pattern of intensification noted for the period 1970–84? The very general and declining quality of CIDT data fail to capture the diversity of peasant coping strategies. These dynamics are discussed more fully below with reference to the case of Katiali. However, the cotton company data do indicate a general trend that was also observed in Katiali: cotton area per grower increased at the same time that farm input levels (e.g. the amount and number of fertilizer and pesticide applications) declined over this period. That is, a common response to the worsening economic situation was an *extensification of production* as farmers decided to spread their more costly inputs more thinly over a larger area. This extensification strategy is congruent with Bernstein's theory of the "simple reproduction squeeze" which hypothesizes that as the terms of trade worsen, peasants will either reduce consumption levels or intensify production (or both simultaneously) in order to maintain previous income levels. I argue in this chapter that this "simple reproduction 'squeeze' . . . summarized in terms of increasing costs of production/decreasing returns to labor" was both experienced and resisted by peasant farmers in northern Côte d'Ivoire.[10] Through a variety of coping strategies, peasants sought to maintain income levels by simultaneously reducing the costs of production and increasing returns to labor. The data for neighboring countries suggests that this extensification strategy was not simply confined to the cotton growing areas of northern Côte d'Ivoire.[11] We now turn to the case of Katiali where extensification was, in fact, just one of many strategies pursued by peasant farmers to deal with the downturn of the cotton economy.

The extensification of cotton

There are two dimensions to extensification. The first is an expansion in area cultivated per grower; the second is a reduction in the level of inputs (capital, labor) per unit area. These two facets of extensification were evident in Katiali over the period 1985–95. After the fertilizer subsidy was eliminated in 1984, cotton growers decided to spread this input over a larger than recommended area. Table 6.2 provides measures of both types of extensification by comparing the area in cotton declared by the members of 38 households in Katiali with the area recorded by the CIDT for the same households. The difference ranges from 4% to 36% of the cotton area of Katiali. This undeclared area demonstrates that input rates recommended by CIDT were not being followed at the village level. A measure of this under-dosage can be obtained by dividing the undeclared area by the CIDT-declared area. The under-dosage rate reached 57% in 1988 in Katiali. That is, the amount of inputs (e.g. fertilizer) applied to cotton fields was 57% less than the recommended levels because they were spread over a larger area. The objective of farmers pursuing the extensification strategy is

Table 6.2 *Undeclared cotton area and rates of extensification: Katiali*

Year	Survey data Katiali (ha)[1]	CIDT data Katiali (ha)[1]	Undeclared area (%)	Under-dosage rate (%)
1981	73.25	70.50	+4	4
1985	79.50	58.50	+26	36
1988	153.75	97.75	+36	57
1994	86.75	60.85	+30	42

Source: Field data; [1]N = 38 households

to increase incomes by reducing the costs of each kilogram produced. This appears to be a rational strategy. Research in Mali's cotton growing areas shows that although yields per hectare drop under extensification, net incomes tend to increase.[12] In Côte d'Ivoire, the ability of cotton growers to manipulate input rates and cotton area was facilitated by the restructuring of the cotton extension system so that CIDT monitors began to meet with groups of cotton growers in one location rather than with individual farmers in their fields. Thus extension agents were unable to monitor the production process as closely as they had in the past. In the end, the extensification strategy was unsuccessful (farmer net incomes fell to historic lows) because of the poor performance of the GL7 variety under reduced input conditions.

The under-dosage measure most likely underestimates the degree of extensification, for it assumes that farmers were buying the recommended amounts of fertilizer (NPK and urea) for the cotton area they did declare.[13] Even CIDT's data indicate this was not the case. Figure 6.3 shows that the amount of fertilizer applied to the cotton area recorded by extension agents declined significantly since the early 1980s. From a high of 243 kg/hectare in 1981–82, the amount of fertilizer purchased for (*but not necessarily applied to*) cotton fell to 180 kg between 1993–95 – a 25% decline. It is common knowledge among farmers and extension agents alike that cotton growers divert fertilizers from cotton to food crop fields, as well as sell them to other farmers.[14] This fertilizer under-dosage is, in fact, a common practice throughout the cotton-growing areas of West Africa.[15]

It is also common knowledge that a cross-border traffic in pesticides has flourished for many years. Up until 1994, pesticides originating from Côte d'Ivoire regularly made their way to the neighboring countries of Burkina Faso and Mali, where similar HYV packages were cultivated but not subsidized.[16] Ivorian cotton growers also sold pesticides locally to Fulbé cattle owners, who used them to rid their animals of ticks. Pesticide sales meant that some farmers were not following CIDT recommended dosages. The

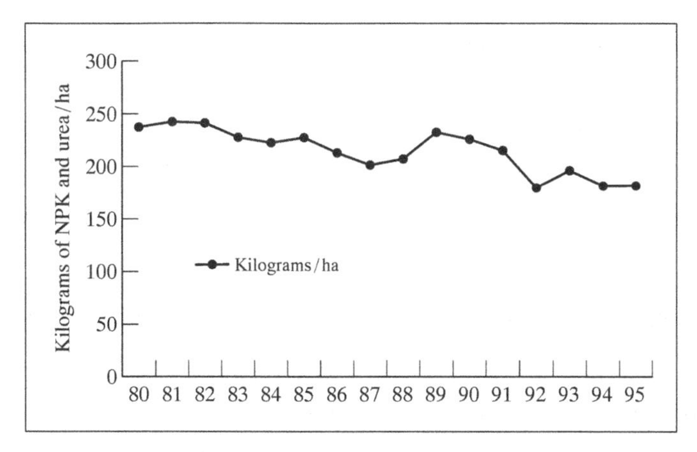

Fig. 6.3 Fertilizer use (NPK, urea) per hectare, 1980–1995, Côte d'Ivoire (*Source:* Field data and CIDT)

impact of such agricultural practices on cotton yields depends in part on the tolerance of the variety cultivated. The more hardy ISA variety better tolerated these less than ideal agronomic conditions. GL7 yields plummeted if regularly scheduled sprayings and doses were not respected. The poor cotton harvest of 1994 is largely attributed to irregular pesticide applications. CIDT noted in its 1994 annual report that "crop pest protection dropped sharply. With the suppression of free insecticides for cotton, farmers reduced the number of sprayings and in certain cases the recommended doses were not respected."[17]

In summary, a process of extensification of cotton began in the second half of the 1980s, as documented in the high percentage of undeclared cotton area and widespread under-dosage of fertilizers and pesticides. The strategy appears rational in light of the increasingly adverse market conditions, especially for lower-income households short on capital and labor. The magnitude of this undeclared cotton area has important implications for the quality of CIDT's data. The Katiali data suggest that CIDT's annual reports for the second half of the 1980s underestimate the number of growers and area in cotton and thus overestimate yields and net incomes. Given the centrality of these parameters in measuring agricultural performance, one might conclude that the cotton revolution was short-lived. However, unprecedented production levels during the second half of the 1990s suggest that the years 1985–95 were instead a major transition period, marked by the dismantling of the CFDT system and the rise of a new cotton economy, in which tens of thousands of small farmers came to hold a new stake in the expansion of cotton.

Contested cropping

Women began to grow cotton in increasing numbers during the second half of the 1980s. By 1988 cotton had become the most important cash crop for women in a third of the 38 households surveyed in Katiali who cultivated, on average, a half a hectare. Most women were not registered with CIDT nor with the local production and marketing cooperative to receive subsidized inputs and credit. They most often had to obtain seed, fertilizer, and pesticides from individuals willing to sell them. This informal, if not invisible, nature of women's cotton growing was made possible by the reduced surveillance of cotton monitors under the new extension system, the emergence of an ox-plow rental market, and the informal marketing of cotton inputs. Women's cotton fields were usually planted late because they had to wait until the fields of ox-plow owners were prepared. Since input levels were also generally insufficient, yields were correspondingly low (396 kg/ha). Nevertheless, women were attracted to cotton growing because they could obtain relatively large sums of money at one time in contrast to the meager earnings gained from selling small quantities of rice or peanuts in weekly village markets.[18]

Despite the adverse agronomic and macro-economic conditions of cotton growing in the early 1990s, some women still sought to grow this single most important cash crop. However, their efforts to do so were increasingly thwarted by their husbands and other men. The results of a survey of 38 households undertaken in Katiali in 1994 showed that the number of women cultivating cotton had declined from 31 to 4 between 1988 and 1994. When asked why they no longer possessed a personal cotton field, more than a quarter (27%) of the 24 women interviewed in 1994 cited conjugal conflicts related to the desire of male household heads to control their labor power. Respondents stated in no uncertain terms that their husbands felt threatened by their gaining a measure of freedom with their cotton incomes. Many women stated that their husbands wished for them to remain subordinate, forcing them to say "excuse me, excuse me, can you spare five centimes. . . ?"[19]

A second group of responses (22%) explained the decline in women's cotton to insufficient labor supplies during critical periods in the agricultural calendar. This was linked not only to limited access to household labor but also to the lack of resources to hire non-household labor. Relatively wealthy women commonly paid for non-household labor to plow their fields in cash. Poorer women more often bartered their labor (or food) to pay for non-household labor. A common arrangement was to barter 10 to 20 days of harvesting cotton in the fields of an ox-team owner for plowing a half-hectare field. In general, Senufo women were more likely to provide in-kind services to obtain farm inputs than Jula women, who more often paid in cash.

A full quarter of the responses centered on the absence of women at the cotton cooperative market. Women said that they had "to hide behind" someone to sell their cotton. This was particularly true for Senufo women who were not registered with the cooperative. They depended on either their husband, son, or some male friend to bring their cotton to the market, sell it for them, and then load it into trailer-truck containers. Women felt that they were not receiving the income that they thought was their due, openly stating that their husbands in particular were keeping some of their earnings.

Finally, 19% of the responses focused on the lack of profit to be made in cotton, citing the GL7's poor performance in their fields, as well as the higher costs of production following the devaluation and end of the pesticide subsidy.

Men complained that their wives were making too many demands on limited household resources and were not spending enough time in household fields. When asked why men like him no longer supported their wives' cotton growing, Tiékundôh Silué responded:

> The women do not have oxen. So we have to plow their fields. It is too much! Moreover, they don't go themselves to the warehouse to get fertilizer. They take it from their husbands. And if they don't make much money from their cotton, they are not happy . . . Today, women want to work for themselves; they want to hire other people to plow their field, to weed it, to harvest it. But when they arrive with their cotton at the market, the young men refuse to load their cotton into the truck. The young men say that their mothers don't work in the family's fields like they did in the past now because they grow cotton. They decided to put an end to this by refusing to load women's cotton [into the truck] . . . If you want to know the truth, women show less respect to their husbands. They want to grow rice, peanuts, cotton, and then work with their husbands. But can they really work on a regular basis in their husband's fields?[20]

Tiékundôh went on to say that in the past, women only worked two days a week in their fields – *koundièle and kong*. For the past three to four years, he said, they work in their fields whenever they want. Senufo men "noticed" that their wives were not going to work in household fields on Friday, a day when many men took a rest day for usually religious reasons,[21] nor on Saturday, the day of the weekly market in Katiali. When combined with women's customary work days (*koundièle* and *kong*) for their personal fields, men "discovered" that their wives were only working two to three days in six to seven-day weeks.

Katienen'golo Silué, an influential Muslim Senufo with five wives, decided to take action. One day in 1995, he simply informed his spouses that they could work in their personal fields on Friday and Saturday, but no longer on *koundièle* and *kong*. Other Senufo men soon followed his example. Almost overnight, women's work days were shifted from the Senufo six-day week to the seven-day week calendar. By transferring women's work days to Friday and Saturday, Senufo men were able to gain an extra two weeks of women's work in household fields during the rainy season. Although the math was simple, the cultural implications were profound. In a six-day week, women had ten days per thirty-one day month to work in their individual fields. In a seven-day week calendar, they only have 8 days. Thus, household heads stood to gain two extra work days per month from each of their wives. When totaled over the length of the agricultural calendar (June–December), individual women lost fourteen days of work in their individual fields. The exception to this rule applied to senior wives, who were not subject to the same calculus. Although always contingent upon the degree of spousal harmony, household heads normally allowed their first wives, particularly those who were beyond their childbearing years, to work in their personal fields whenever they chose. The new ruling mainly affected junior wives. Thus, Katienne'ngolo Silué managed to obtain an additional fifty-six days of work (four × fourteen days) from four of his five spouses in household fields. The politics surrounding male household head's attempts to restrict women's cotton growing illustrates what Judy Carney, Michael Watts, and Donald Moore have demonstrated in different contexts. That is, struggles over access to productive resources commonly take the form of "cultural politics" in which the content and meaning of cultural categories such as women's individual field days within customary calendars are restructured and redefined.[22] The disassociation of women's personal field work days from the Senufo six-day per week calendar is a further example of the gradual erosion of the Senufo culture of production that had been taking place since the 1920s, when the *segbo* began to break up (Fig. 4.2).

In summary, the cotton boom of the 1970s and 1980s opened up new opportunities for women to expand their incomes and gain a measure of economic autonomy within the household. The availability of ox-plow rental services and farm inputs obtained in informal markets allowed them to cultivate small cotton fields (one quarter to half hectare). The worsening economic environment, resulting from the institutional reforms of the 1990s pushed through under structural adjustment and the high profit taking by CIDT (see below), created tensions within households over access to productive resources, notably labor. Contestation between men and women at the intra-household level took the form of circumscribing women's cotton growing and redefining, through local cultural categories, the days of the week when women could work in their personal fields. The

Table 6.3 *Indebted cotton growers: Katiali, 1989–95*

	Sample		Village	
Year	Number	Percent	Number	Percent
1989/90	2/38	5	17/170	10
1990/91	5/38	13	60/193	31
1991/92	19/38	50	57/157	36
1992/93	13/38	34	66/150	44
1993/94	10/38	26	41/148	28
1994/95	14/38	37	54/121	45
1995/96	16/34	47	42/123	34

Source: Household surveys (N= 38); GVC and COOPAG-CI, Katiali

following sections on the impact of household debt on the evolution of community level institutions (e.g. village cooperatives) and the role of federated farmer organizations in negotiations with CIDT over the distribution of cotton revenues takes us to additional levels in the multi-layered negotiations driving this agricultural and social history.

Managing debt

Defaulting on loan repayments, while minimizing personal economic losses, was another method used by households to cope with adverse economic and agronomic conditions. Table 6.3 shows that a large percentage of cotton-growing households in Katiali failed to reimburse input suppliers over the period 1989–95. Indebtedness results from the inability or unwillingness of growers to reimburse input supply creditors when they sell their cotton. Up until 1986, CIDT dealt directly with cotton growers to settle debit accounts. Beginning in 1987 the cotton company shifted responsibility for loan repayments to village-level marketing cooperatives (GVC). If individual cotton growers failed to make their payments, the cooperative had to reimburse CIDT completely for fertilizer deliveries. To recuperate unpaid loans, CIDT simply retained part of the commission (*ristourne*) it owed to cooperatives for their marketing and input supply services.[23] The case of the GVC of Katiali for the year 1990/91 is illustrative. The GVC was to receive 2,494,000 FCFA ($9,000) in commission for its production and marketing services, but CIDT deducted 1,495,000 FCFA ($5,436) because 45 cotton growers (20% of all GVC members) were unable to reimburse the company for its fertilizer advances. The GVC also had to reimburse the Swiss chemical company, Ciba-Geigy, 887,000 FCFA ($3,225) for herbicide advances because 44 cooperative members failed to make their payments.

Table 6.4 *Percentage of households in debt by income group: Katiali, 1991–94*

Income group	Number in sample	1991	1992	1993	1994
I	8	50	25	13	13
II	22	64	45	36	50
III	8	13	13	13	25
	Average	50	34	26	37

Source: Field data; N = 38 households

After these two deductions from the GVC's marketing commission, all that remained was 112,000 FCFA ($407).[24] GVC members still owed 645,000 CFA ($2,345) to the herbicide manufacturer Rhone-Poulenc and 208,000 FCFA ($756) to the National Agricultural Development Bank for ox-plow loans. The likelihood of recuperating these losses during the 1991/92 season was dimmed by the prospects of an exceptionally low cotton harvest.[25]

The point worth emphasizing here is that the cooperative structure permits a "free-rider" situation in which the collective, not the individual, is penalized for defaults on personal loans. In the short term, this allows a household head to escape the hardships associated with the simple reproduction squeeze by such methods as crop mortgaging or selling household food crops immediately after harvest.[26] Chronic defaulters manage their debt by selling only a portion of their crop under their personal name to the cooperative. The amount delivered is usually not sufficient to erase their running debt. To make sure that they obtain some cash payment, free-riders sell most of their cotton under the name of a friend or relative who simply pays him/her after the cooperative pays its members. As the number of producer cooperatives grew in the 1990s, some farmers took advantage of the possibility of obtaining input credits from one cooperative and selling their cotton to a different one, thus avoiding repaying their debts. This was true of most cotton growers at this time, regardless of their social standing. Table 6.4 shows that indebtedness was widely distributed across income groups for the years 1991–94. There were 56 cases of indebtedness for the 38 households surveyed. The table indicates the percentage of indebted households by income group. For example, 4 of the 8 upper-income households in 1991 were in debt (50%). Middle-income households bore the brunt of the debt. This group accounted for forty-three of the fifty-six cases – 77% of the total. Jula households dominated the debt lists, representing 63% of indebted households. Only middle-income Senufo households fell

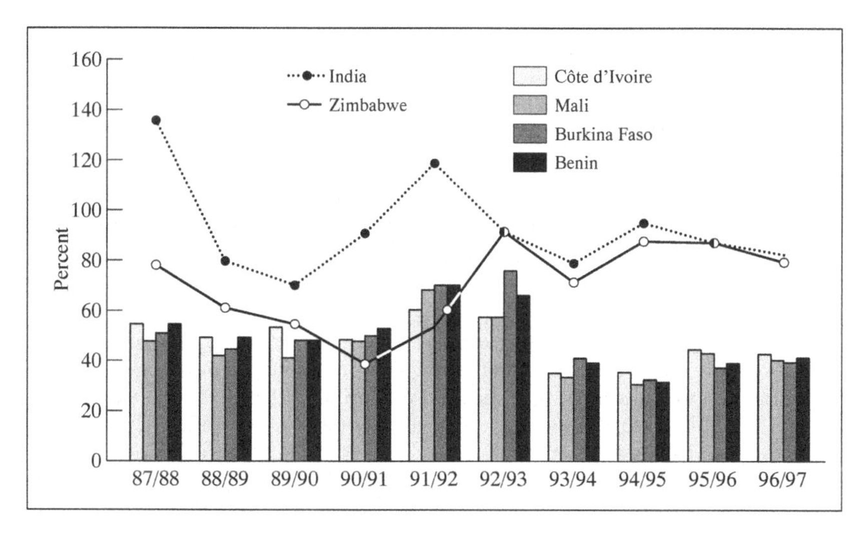

Fig. 6.4 Cotton growers' share of world market price (FOB) for selected countries (*Source:* Amprou, *et al.*, "Le coton")

into debt. There were no upper- nor lower-income Senufo households in debt to the cooperative over this period.

The chronic nature of peasant indebtedness is in large part related to the extraordinarily low share of world market prices they were receiving for their crop. Figure 6.4 reveals that throughout the late 1980s and 1990s, West African cotton growers typically received less than half of the FOB price for cotton lint. The main beneficiaries of peasant cotton were CIDT, CFDT and its cotton trading subsidiary, COPACO, and the Ivorian textile industry, which was obtaining subsidized cotton during these years. In contrast to India and Zimbabwe, West African cotton growers were consistently underpaid for their labor product. Their efforts to make cotton work agronomically and commercially did not benefit them as much as it should have. Within the CFDT/CIDT system, which operated under the ideologically laden banners of agricultural and regional development, peasant contributions to the cotton revolution ultimately benefited Ivorian and French industrial and commercial firms. Farmers were not oblivious to the dynamics of capital accumulation occurring at their expense. Their use of the cooperative structure to strike back at CIDT was an innovative strategy aimed at capturing a larger share of the world market price for cotton. In light of the importance of village-level production and marketing cooperatives in peasant coping strategies, it is worth looking more closely at their origins and recent evolution.

The cooperative movement

The current cooperative movement dates from the mid-1970s when cotton growers successfully replaced CFDT/CIDT extension agents in village-level cotton marketing. According to Adama Koné, the accountant of the GVC of Katiali between 1975–80, the cooperative movement was spurred by grower suspicions that CFDT/CIDT agents were engaging in fraudulent practices (e.g. rigged scales, dishonest bookkeeping) in local cotton markets.

> Peasants thought that CFDT was stealing lots of money from them in the early 1970s. People grew cotton but didn't know the weight, didn't know what the agent wrote in his book, and then they received money. They believed they were losing money. The DMC (Direction de la Mutualité et de la Coopération) came to the village and held a meeting with the Secretaries of the PDCI (Parti Democratique de la Côte d'Ivoire) and the quarter and village chiefs. The DMC told them that "your sons will buy your cotton and tell you how it all works. You need to give to the GVC your sons who know how to read and write."
>
> After one to two years, tensions grew between CIDT and the GVC because CIDT thought we were against them. They no longer provided us with an adding machine and supplies. "You have to manage on your own," the [CIDT] zone chief told us.[27]

With the assistance of the Direction de la Mutualité et de la Coopération (DMC) of the Ministry of Agriculture, farmers organized themselves into marketing cooperatives (Groupement à Vocation Coopérative or GVC) to collect cotton at the village level.[28] This primary stage of marketing involves the weighing and recording of each cooperative member's harvest, loading the seed cotton into large shipping containers, and eventually paying growers the value of their product minus loan payments.

In return for its marketing services, the GVC receives a commission (*ristourne*) for each ton of seed cotton collected.[29] These funds have been used to co-finance social infrastructure projects such as primary schools, clinics, and water pumps in rural communities. In 1975, 13 cooperatives marketed 2,462 tons of cotton. In 1982 there were 192 cooperatives responsible for 52% of total production. By 1989, 99% of all cotton produced in the country was managed by 634 cooperatives, representing more than 130,000 farmers residing in some 3500 villages.[30] The GVCs reportedly received more than 1 billion FCFA ($3,672,000) in payment for their services that year.[31] The most experienced GVCs also function as production cooperatives by organizing the distribution of inputs (fertilizers, herbicides) for both cotton and food crops to its members. The credit-in-kind system allows group members to obtain inputs at the beginning of the rainy season and to reimburse the GVC at market time when loans are deducted from

Plate 20 Weighing cotton at the GVC cooperative scale, Katiali, January 1981

Plate 21 GVC treasurer distributing cotton earnings to cooperative members, March 1982

cotton earnings. Credit for the purchase of oxen and plows was also available to GVC members through the National Agricultural Development Bank (Banque Nationale de Développement Agricole) up until 1991.[32] Loans were typically given to cotton growers with good performance records. The repayment period was normally three years at an interest rate of 12%.

Beginning in 1982, some cooperatives organized themselves into district-wide unions called Union des Groupements à Vocation Coopérative (UGVC), in order to represent more effectively the interests of member cooperatives in their dealings with agricultural input suppliers, CIDT, DMC, and other agencies. Like the GVC, the leadership of the cooperative unions received training from the DMC. In 1992, there were twenty-two unions representing 40% of the GVC in the cotton-growing areas.[33] The unions and individual GVC joined to form a federation in 1991. The new farmer organization, called l'Union régionale des enterprises coopératives de la zone des savanes de Côte d'Ivoire (URECOS-CI) (Regional Union of Cooperative Enterprises of the Côte d'Ivoire Savanna Zone), sought to coordinate the activities and promote the interests of individual cooperatives and unions around agricultural policy issues, particularly producer prices.[34] URECOS-CI administrators received training from the International Labor Office of the United Nations. The federation also had close ties to the Parti Démocratique de la Côte d'Ivoire (PDCI), the ruling political party.[35] This evolving organizational structure became a springboard for the cotton grower mobilization that took CIDT by surprise during the 1991/92 harvest period.

Striking cotton markets

As early as 1989, when CIDT reduced producer prices for second-grade cotton (from 105 to 100 FCFA), representatives of the GVC Unions began to talk about organizing a market strike.[36] Tensions between growers and the cotton company continued to build as net incomes steadily declined. Falling incomes were linked not only to reduced prices and inflation but also to the poor performance of GL7 in peasant farming systems. In 1991, 70% of the cotton area in the north was planted in the highly demanding glandless variety. Yields per hectare were the lowest recorded by CIDT since 1975. Considering the level of extensification, it is safe to assume that average yields were even lower. Moreover, for the previous three years, CIDT graded a large proportion of peasant cotton as low quality, thus reducing incomes even further. In short, after nearly three decades of nominal price stability and periodic increases, growers were now witnessing the breakdown of the "CFDT system." The Agricultural Price Stabilization

Fund (Caistab) was heavily indebted to export agents, and late payments to producers were becoming the norm.[37] At the same time, the World Bank was placing pressure on the Ivorian government to adopt economic liberalization policies, one of which was to disengage from private sector activities. As a result, CIDT was restructured in 1991 under the Contract Plan ("Convention-cadre") negotiated between CIDT and Ivorian government. Under the terms of the agreement, the state was to disengage itself from company operations and CIDT was to operate more like a private company. More importantly for the future of CIDT and the "CFDT system," the government also decided during the Contract Plan negotiations that CIDT would eventually be privatized.[38] CIDT began to manage its own "stabilization fund" (Fonds de Garantie de la Filière Coton) to which producers, CIDT, and the state contributed funds. A special tripartite committee was also created to set producer prices. It was comprised of six members representing three groups: two from the Ministry of Agriculture, two from CIDT, and two representing cotton grower organizations.[39] A simple majority of votes was sufficient to set prices.

When the tripartite committee met to set the 1991 prices, CIDT reportedly made little attempt to negotiate a price acceptable to all committee members. It recommended that seed-cotton prices be reduced by 10% and that input prices be increased. Over the objections of the two cotton grower representatives, CIDT's proposal was approved by a majority vote. Declaring that "CIDT has gone too far with its swindling,"[40] the cotton grower representatives returned to Korhogo and mobilized an unprecedented resistance campaign through the evolving producer cooperative structure.[41] During the first months of the cotton marketing period, they held meetings with GVC and union leaders to urge them to withhold cotton from the market. The "cotton strike," as it became popularly known, effectively shut down all of CIDT's cotton gins in November and much of December. Trucks circulated throughout the cotton-growing areas leaving containers behind for cooperatives to fill. When they returned to pick them up, they found them empty.

The strike was called off in late December only after the producer price committee met again and reached an agreement to increase the purchase price of second-grade cotton from 75 to 80 FCFA/kg and to reduce the price of NPK fertilizer (10% below the 1990 level) for the 1991 and 1992 seasons.[42] The committee also agreed to raise the urea purchase price by 20% for 1991–92 but this was still 20% less than CIDT's originally recommended price. More importantly, the cotton market strike was an unprecedented display of peasant political power that effectively challenged CIDT's control of the vertically integrated cotton sector. The success of the cotton strike sent a signal to CIDT that it could no longer dictate the terms of its "contract" with growers but had to negotiate with them in good faith and in a transparent manner.[43] The strike marked a turning point in how business was conducted between CIDT and cotton growers.

Two years later (1993), farmers again refused to load CIDT trucks after the company announced (at harvest time!) that purchase prices would be reduced once again. While negotiations took place between grower representatives and CIDT, cotton gins stood idle for nearly two months.[44] With the devaluation of the FCFA in the offing, the company finally agreed to increase producer prices, albeit at a modest level. In absolute terms, more than $10 million was distributed to cotton growers after the price increase. However, the producer's proportion of the total value added in the cotton sector dropped from 25% to 15%. Some of the main beneficiaries of the devaluation were the local textile firms – Gonfreville, Cotivo, and Utexi. In 1994 the Ivorian government (the majority shareholder of CIDT) instructed CIDT to sell cotton lint to these companies below its cost price.[45] The $6 million subsidy reaped by these textile firms was the equivalent of 12 FCFA per kg of seed cotton.[46]

Pressured by cotton growers on one side and by the World Bank, which was critical of CIDT's paltry post-devaluation producer price,[47] on the other, CIDT raised the purchase price of top-grade cotton to 150 FCFA for the 1994 season. But with inflation running at 35% and pesticides costing dearly, farmers refused to sell at that price. Beginning in November 1994 village-level cotton markets were idle and CIDT's ten cotton gins ran well below capacity, if they functioned at all. It was not until early January 1995 that CIDT finally capitulated to peasant resistance and raised the producer price by 10 FCFA.[48] Cotton growers were not the only winners. Ivorian textile firms successfully lobbied the Ivorian government to force CIDT to hand over a $12 million subsidy in the form of locally produced cotton lint.[49] CIDT's general manager pointed to the contradiction in government policy when he said in an interview to a local newspaper: "We told the government that you cannot ask us to operate efficiently and at the same time subsidize local industry."[50]

The 1991 cotton agreement (Contract Plan) between CIDT and the state supposedly gave the cotton company absolute control over cotton marketing. Explaining why the government failed to implement this reform takes us to the patrimonial heart of the Ivorian political system. Demery describes the basic political-economic dynamic:

> During times of recession, political control over the economy becomes even more important: mechanisms that enhance rent-seeking and patronage become critical to maintaining political support . . . Any attempt to loosen the grip of the government on the functioning of the economy (in accordance with the free market paradigm) is thus likely to be inconsistent with the political economy, and thus to be resisted.[51]

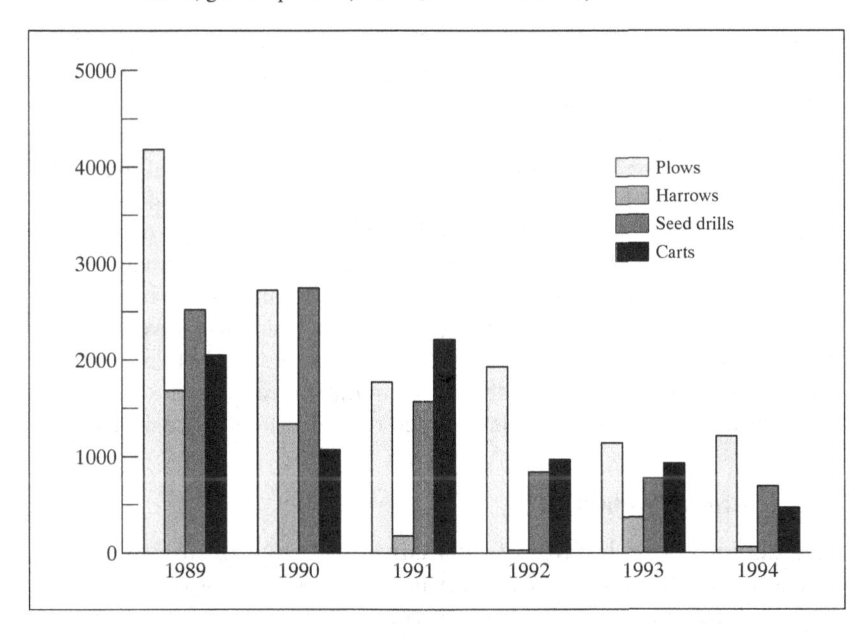

Fig. 6.5 Evolution of animal traction equipment sales, Côte d'Ivoire, 1989–94 (*Source:* CIDT)

Despite farmer struggles to lighten the burdens of structural adjustment in the cotton sector, their ability to participate in the capital and labor intensive HYV program became increasingly constrained. The Côte d'Ivoire Living Standards Survey indicated that poverty rates rose significantly in rural areas during the second half of the 1980s. As net incomes continued to fall in the first half of the 1990s, investments in labor-saving and yield-enhancing technologies conspicuously began to decline.

The end of modernization

CIDT data on animal traction equipment sales suggest that farmers found it increasingly difficult to invest in new agricultural technologies (Figure 6.5). The company's data show an overall decline in the number of ox-plows, harrows, seed drills, and carts purchased by farmers between 1989 and 1994.[52] Ox-plow sales dropped precipitously from over 4,000 in 1989 to around 1,000 in 1993 and 1994. This brake on agricultural modernization suggests that structural adjustment in the cotton sector ushered in a recessionary phase into the rural economy.

The importance of animal traction to the cotton program is evident in the fact that almost half (48%) of the cotton area is farmed by ox-plow households.[53] Indeed, there is a strong relationship between the expansion in cotton area with the diffusion of ox-plows. The greatest increase comes when peasants switch from manual cultivation to animal traction. The second largest increase stems from the incremental expansion of area cultivated by ox-plow households. Yet less than a quarter (23%) of cotton growers in 1994/95 used animal traction. The majority of growers depended on manual labor. Ox-plows have also been key elements in the process of extensification discussed above. As the purchase of animal traction equipment became economically difficult, this crop diversification strategy became less of an option for cotton growers seeking to mitigate the worsening terms of trade. Under these circumstances, it is not surprising that peasants began to invest their limited resources in other cash crops.

Crop diversification

Abandoning cotton in favor of alternative cash crops or investments has been a classic strategy of peasant farmers in the history of cotton in Côte d'Ivoire. Farmers have traditionally turned to food crops when prices were comparatively more attractive. The CFA franc devaluation of 1994 was a stimulus to food production, as grain imports became more expensive and demand increased for domestic food supplies.[54] Such a trend is apparent in the case of Katiali where cotton area declined relative to food crops (Fig. 6.6) over the period 1988–94. From a high of 47% of the total cultivated area in 1988, cotton area declined to 41% in 1991 and then to 34% in 1994. The area in rice and maize expanded over the same period. Women cotton growers in particular shifted to rain-fed rice after they found it unprofitable to cultivate the highly demanding GL7 variety. The number of women growing cotton declined as rapidly as it had risen in my sample of 38 households. From a high of 31 women cotton growers in 1988, just 4 women continued to grow cotton in 1994.[55]

At the same time that farmers began to place greater emphasis on food crops, they also began to look for alternative cash crops. Two crops of particular interest were cashews and mangoes. As early as 1988, farmers began to plant trees in their food crop fields. After these fields were put into fallow, the trees were pruned and protected from bush fires. Figure 6.7 shows the area in cashew and mango orchards for 1995. Up until that year, cashews were bought by both local and regional merchants based in Korhogo. When a new cooperative formed in the Niofoin area in 1996 that specialized in cashew marketing,[56] Katiali growers sold over three tons in its first year.

Rather than abandon cotton outright, some farmers pressured CIDT to

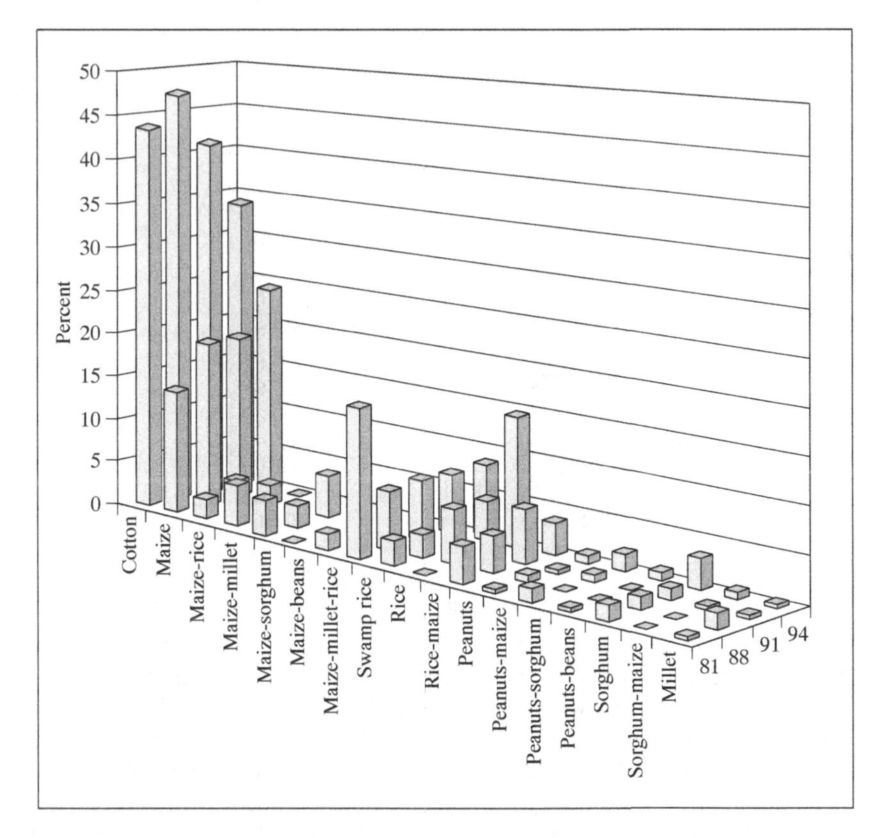

Fig. 6.6 Crop area for selected years, Katiali (*Source:* Field data)

abandon glandless cotton varieties and return to the more flexible ISA variety. For example, cotton growers in the Ferké area refused to plant glandless cotton in 1992–93.[57] Côte d'Ivoire cotton growers were not alone in rejecting GL7. According to CFDT sources, groups of cotton growers in Mali, Burkina Faso, Bénin, and Togo also refused to plant GL7.[58]

For upper-income households, investing in livestock proved to be an attractive diversification option, especially between the 1980s and early 1990s. Between 1982 and 1992, the number of sedentary cattle nearly tripled in Katiali, rising from 578 to 1,417 head. This was in addition to the 380 oxen owned by Katiali's farmers in 1992. Dwarfing these numbers was the spectacular increase in Fulbé transhumant cattle, whose numbers grew from 3,310 in 1982 to 7,033 in 1992. The growth in Fulbé cattle was mainly due to the immigration of new herders into the region. Crop damage caused by Fulbé cattle in farmers' fields was a major source of tension

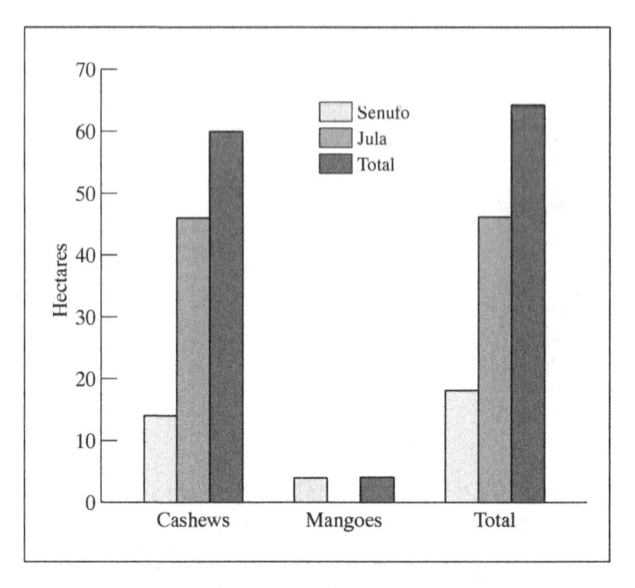

Fig. 6.7 Area in orchards, Katiali, 1995 (*Source:* Field data)

between farmers and herders.[59] Notwithstanding the sometimes bitter conflicts dividing the two communities, Fulbé herds were the principal source of cattle bought by farmers. Figure 6.8 shows a steady rise of oxen ownership between 1982 and 1990 for Katiali. However, a notable decline took place after 1990 which the CIDT extension agent explained as the result of either oxen sales, or deaths, or farmers deciding not to grow cotton. In the latter case, oxen were not counted by the agent.

Cooperative turns

The cotton market strikes and restructuring of CFDT in the early 1990s demonstrated to peasants and company officials alike that the 30-year-old parastatal-regulated model of cotton development was in a profound state of transition. Cotton farmers not only spoke in a forceful voice by virtue of their presence on the tripartite committee but, more importantly, they now possessed the organizational strength to speak even more loudly through collective action. The scope of cooperative activities broadened (e.g. input distribution, credit recovery) as CIDT was required under the World Bank-brokered Contract Plan to devolve some of its former commercial activities to the private sector. The organizational structure of GVCs also evolved with the formation of the regional-level farmer organization, URECOS-CI. A second regional farmer organization emerged in late 1992 that

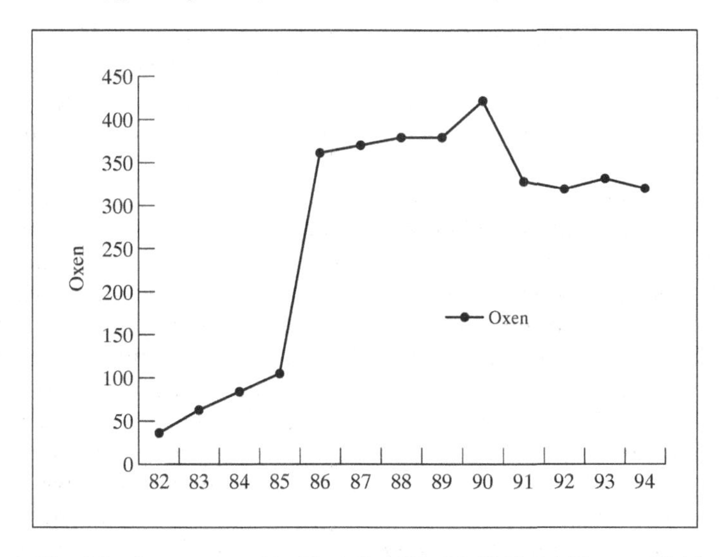

Fig. 6.8 Oxen ownership, Katiali, 1982–86, 1988–94 (*Source:* Field data)

appealed to farmers seeking to address the free-rider problem plaguing the cooperative structure. The Farmers' Cooperative of Côte d'Ivoire (La Coopérative des Agriculteurs de Côte d'Ivoire or COOPAG-CI) began to build its membership in 1993–94 by promising to exclude repeat loan defaulters, to reduce the price of inputs by eliminating intermediaries, and to make quick payments to farmers. It was popularly viewed as an "opposition party" cooperative (associated with the Front Populaire Ivoirien or FPI) in this pro-democracy period during which multi-party politics began to manifest itself in rural communities.[60] Its ability to deliver on all three pledges resulted in a large number of cotton growers switching from the GVC/URECOS-CI structure to the COOPAG-CI. A third farmer organization appeared in 1995 called the Cooperative of Côte d'Ivoire Farmers (La Coopérative des Exploitants Agricoles de Côte d'Ivoire or CEACI) which competed with URECOS-CI and COOPAG-CI.[61]

This new competition for cooperative members and market commissions forced GVCs to address the free-rider problem.[62] Some ejected chronic defaulters from their roles,[63] but for many members, this action did not go far enough. In communities like Katiali, cotton growers began to split from their parent GVC to form smaller cooperatives, sometimes at the residential quarter level. They declared that "good farmers" (i.e. those who did not default on their loans) would form the core of the new cooperatives.[64] The multiplication of producer associations reflected, on the one hand, a growing social differentiation between prosperous or "good" farmers and poor, chronic defaulters. However, the generalized level of poverty in the community, and the fact that even good farmers default in bad years, suggest that additional factors were behind this splintering of the GVC structure.[65]

The proliferation of producer organizations was partly linked to the desire of different communities to capture the supplementary payments (*ristourne*) and commissions received by cooperatives for marketing cotton and distributing inputs such as fertilizers.[66] The GVC of Katiali, which formerly included growers residing in the smaller neighboring communities of Gbodonon, Komon, and Lygnoublé, used its supplementary payments to invest in social infrastructure such as the village elementary school, clinic, and cooperative warehouse. In the late 1980s and early 1990s, Katiali's neighbors formed their own cooperatives in part to control and invest marketing commissions and supplementary payments in their own communities. The fissioning of the GVC structure and the emergence of new farmer organizations like COOPAG-CI were also linked in some cases to farmer discontent with the opaque management style of GVC leaders.[67] Suspicion of corruption, late payments to growers, and free-riding were central concerns of disaffected cotton growers seeking better services, higher incomes, and better technical support.

Finally, the pace of institutional change taking place at the local and regional level must also be located within the larger processes of political and economic liberalization taking place in the context of the Ivorian fiscal crises of the 1980s and 1990s.[68] Specifically, the emergence of multi-party politics in 1990 allowed for a diversity of views and institutions to flourish. Prior to these political reforms, dissent was stifled by the one-party political machine of the PDCI. In the new political space, opposition parties and interest groups were quick to speak out and challenge government policies, although some still feared reprisal.[69] The involvement of political parties in the creation of farmer organizations speaks to the importance of political party competition in agrarian politics.

Economic liberalization also created openings for farmers to redefine their relationship with CFDT and the Ivorian state as well as influence agricultural policies. Burdened with one of the highest per capita debts in the world, the Côte d'Ivoire government was forced to accept a series of IMF and World Bank structural adjustment programs during the 1980s and 1990s as a condition for continued financial assistance. Although the neo-liberal economic policies did not result in the "privatization of everything,"[70] the Ivorian government's privatization policies did lead to the progressive dismantling of CIDT and the building up of farmer cooperatives. Indeed, one of the major missions of the Agence nationale d'appui au développement rural (ANADER) (National Rural Development Support Agency), which was created in 1993, is to promote producer associations at the village level (GVC and GVC Unions) and thus contribute to the development of professional agricultural organizations. Driven by rural producers, aid donors, and the state, these institutional developments restructured the rural economy and created new arenas in which peasant farmers could participate in the making of national agricultural policy.

In summary, over the course of five years (1991–95), an invigorated peasant farmer cooperative movement profoundly altered the conditions under which CIDT operated. Although the company still retained its monopoly in national cotton markets by virtue of owning all of the country's ten cotton gins, cotton growers began to exert greater influence on the setting of prices and on the timing of cotton deliveries, and thus on gin operations. The privatization of CIDT, a condition of the World Bank's $100 million loan to fund Côte d'Ivoire's agricultural sector adjustment program, promised to further undermine the CFDT system. The French company lobbied vigorously against the breakup of the Ivorian cotton company, in which it had a 30% capital share at the time. It hoped to avoid setting a precedent for the dismantling of other franc zone cotton companies in which it was also an important minority shareholder.[71] Company officials sought to persuade the World Bank that the future of the West African cotton sector and farmer welfare were dependent on the CFDT system.[72] François Béroud, CFDT's director of rural development, argued that peasant farmers were dependent upon its vertically integrated structure to ensure access to credit, inputs, and equipment. If this structure was allowed to be dismantled, he stated, the cotton sector would fall apart and rural livelihoods would be jeopardized. The fact that cotton farmer organizations were keen to break up CIDT's monopsony, and that CFDT had a major financial stake in maintaining its presence in franc zone cotton growing, was left out of their story.[73] In a move indicative of cotton grower interest in obtaining a larger slice of the agrarian pie, the three regional cooperative groups (URECOS-CI, COOPAG-CI, and CEACI) joined forces with two foreign agribusiness firms (Continental Eagle and the Commonwealth Development Corporation) to form a company (SOFINCO) that bid on three of the six cotton gins put up for sale by the Ivorian government.[74] Although their bid was not competitive, the Ivorian farmer organizations did not go away empty handed. The Côte d'Ivoire state maintained a 30% capital share in the "privatized" gins for a three-year period after which it planned to sell its share to private parties. Farmer organizations reached an agreement with the state to purchase a 10% interest in the privatized gins.[75]

Conclusion

The cotton crisis of 1985–86 led CIDT to undertake a series of reforms that had major repercussions for peasant cotton growing. Prodded by the World Bank to adopt liberal economic policies, the government-controlled parastatal company attempted to reduce its operating costs by slashing producer prices, eliminating fertilizer and pesticide subsidies, and restructuring its extension system. Its support of producer cooperatives in the spheres of marketing, input supply, and credit recovery were equally aimed at reducing operating costs. Simultaneously CIDT sought to increase its profits by promoting a glandless variety of cotton that promised higher gin yields and new market niches. Following the 50% devaluation of the FCFA, the financial health of CIDT improved dramatically. Taken together, these changes in the conditions of production and marketing were experienced by farmers as worsening terms of trade and reduced incomes. I have argued that these changes in the structure of the rural economy induced a new wave of agricultural and socio-cultural changes driven by contestation over access to productive resources at the household, community, and national levels. One of the outcomes of these agrarian dynamics was the reversal of the process of agricultural intensification that had made cotton one of the few success stories in West Africa in the 1970s and early 1980s. The extensification and diversification strategies exhibited by farmers were logical responses to deteriorating economic conditions in the context of local factor scarcities – specifically labor and capital. The unintended consequences of dramatically reduced yields were linked more to the poor fit of GL7 to peasant farming systems than to poor farming practices.

The Katiali case study suggests that CIDT's cotton data seriously underestimated the degree of extensification for the period 1985–95. Consequently, CIDT greatly overestimates average yields and incomes for these years. The data quality issue was of major importance to the late-twentieth-century cotton development discourse in which "the CFDT system" was being challenged by the World Bank, which viewed the monopsonistic power of franc zone cotton companies as impeding economic growth and depressing farmer incomes. Inflated data served to buttress the vertically integrated structure of CIDT against the onslaughts of the proponents of market liberalization by giving the appearance that the cotton sector was thriving, thanks to "the CFDT system."

Just as this system was coming under attack by what company president Michel Fichet called the "ayatollahs of liberalism,"[76] it was also being challenged by cotton grower organizations which assumed unprecedented power in setting administered prices and in cotton marketing. The transformative role of local agents on institutional structures was evident in the rapid evolution of farmer organizations, in the repeated mobilization of tens of thousands of small-scale farmers during the cotton market strikes, and in their new participation in the industrial processing level of the sector ("filière") formerly monopolized by CIDT. Cotton growers were simultaneously shaping new cooperative institutions and reshaping the CFDT system as they struggled amongst themselves (e.g. gender politics) and with the cotton company and the state about the distribution of cotton revenues.

The continuing influence of peasant farmers on agricultural policies and innovations was also evident in the sphere of production. CIDT's decision to reverse the glandless program and refocus the HYV package around the more flexible ISA variety shows how farmer decision making both constrains and orients the company's R and D efforts. Unlike in the 1960s, when CFDT could depend upon the coercive power of newly independent African states to impose its new (*Allen*) varieties where it encountered resistance, in the 1990s the French firm and its African partners had to consider the objectives and resources of peasant cotton growers.[77] Thirty years later, the failure of the GL7 package to take off revealed the limits of directed development. Its rejection also revealed the vulnerability of CIDT to the decisions of farmers "to sow or not to sow," as the general manager of CIDT admitted in a 1995 interview.[78] As the owner of 10 cotton gins with a ginning capacity of 300,000 tons, CIDT had a stake in making cotton growing and marketing as attractive to peasant farmers as possible. Where it was not willing to budge, farmers exerted pressure through their individual farm management decisions as well as through their organizations.

7 Conclusion

> Often illiterate and poor, peasants would not be in the position to allow competition to operate between suppliers and buyers in the event of market liberalization.
>
> François Béroud, Director of Rural Development, CFDT, quoted in S. Dupont, "Le Cameroun ou les dangers d'une libéralisation incontrôlée," *Les Echos* (16–17 February 1996), 51.

> We cannot make any mistakes because for a long time peasants were considered to be illiterate and incapable of managing things; so we want to prove to our detractors that today's producers possess the necessary skills to analyze problems and to find lasting solutions.
>
> Martin Yao, Vice-President, Board of Directors, COOPAG-CI, quoted in *Afrique Agriculture*, 271 (June 1999), p. 27.

Agricultural success stories are rare in the African agricultural development literature of the late twentieth century. When examples do appear (e.g. maize in Zimbabwe, cotton in West Africa), it is important to examine their origins and dynamics and the narratives that give them form and meaning.[1] This book's focus on the cotton revolution in Côte d'Ivoire shows that conventional explanations exaggerate the contributions of foreign development experts in explaining the dynamics of agricultural change. My attempt to construct an alternative social and agricultural history challenges this development narrative by drawing attention to the principal rural African sources of the cotton revolution. The objectives of this concluding chapter are threefold: the first is to summarize the main themes and the basic argument of the book. The second is to return to the question raised in the introduction: in what sense can we call the Ivorian cotton story an agricultural revolution? My third aim is to highlight some of the lessons to be learned from the Ivorian case study. In the eyes of many observers, the cotton revolution was an unlikely development in the seemingly circumscribed worlds of peasant farming and agrarian politics.

Closing the price gap: parallel markets and the origins of the CFDT system

To the chagrin of colonial officials, a large part of peasant cotton was never sold in export markets. Chapter 3 showed how the existence of an indigenous handicraft weaving industry offered cotton growers an alternative and more attractive outlet for their crop. Prices paid by indigenous traders in local markets were more competitive than those offered by French merchant houses. With some striking exceptions (e.g. fairs, forced deliveries, bias towards French merchants), the state generally adopted a free market policy in which producers were free to sell their produce to whomever they wished. When food crop prices were competitive, farmers neglected cotton in favor of other crops like yams and rice. Local-level colonial administrators and agricultural agents frequently commented on the price gap between "official" cotton prices and those offered by local cotton and food crop merchants. This domestic competition for cotton, and the existence of alternative cash crops, induced colonial officials to recommend higher export market prices to capture more of the marketed crop. When this failed, the state fell back on its time-tested and morally bankrupt policy of coercion. With the legal end of forced labor in 1946, policy options shifted to more market-oriented mechanisms.

Colonial officials repeatedly faced major dilemmas which led to inconsistent policies. Whether to suppress Jula merchants and the local weaving industry or adhere to the principle of free trade was one such dilemma. Another was the quandary over whether to recruit large numbers of young Senufo men to work in the south or to let them stay in the north to grow cotton. The dilemma over cultivating cotton as an intercrop in a large number of small fields or as a sole crop in large collective fields was tied to the issues of surveillance and incentives. Collective, single-cropped fields facilitated the task of local administrators, crop monitors, and district guards to monitor the implementation of forced cultivation policies. But this solution led to payments going to village or canton chiefs despite the widely held view that payments to individual farmers would serve as a greater incentive to increase production. Faced with such dilemmas, colonial administrators differed widely in their interpretation of agricultural and economic policies. This resulted in considerable variation in their implementation and, at times, created openings for peasants to elude colonial control.

Administrators, private industry, and commercial interests recognized that peasants would only produce more cotton for export markets if they found it profitable. They saw two options: either cotton prices had to be higher than local prices for both cotton and food crops, or cotton yields had to increase significantly so that farmers might profit, even at relatively low prices, by selling more cotton. The wish on the part of the state, cotton traders, and textile firms to keep seed cotton prices low made the second option more attractive. Consequently, cotton development policies in the post-war period centered around the development of a high-yielding cotton package made possible by the development of new pesticides such as DDT. These policies and initiatives bore fruit. The high input/output *Allen* cotton package developed by IRCT and pushed by CFDT became the basis of the impressive growth rates in cotton production during the 1970s and early 1980s. The point worth emphasizing here is that this land-saving technology was inspired in part by the state's dual desire to keep cotton prices low and at the same time to expand output to levels that would satisfy both local and export markets. That is, the parallel market pressure felt by French textile concerns and colonial administrators forced them to consider alternative policy options, from which they selected the one that best suited their interests.

In addition to influencing post-war research and development directions, the parallel market also inspired agricultural policy makers to intervene in cotton markets to eliminate local competition. In contrast to the stereotyped images of the Jula as wily merchants, and the Senufo as hard working but easily duped peasants,[2] Jula traders had historically offered higher prices to cotton growers than French merchant houses. If the Jula could somehow be excluded from cotton markets, then French traders could capture more peasant cotton at low prices. Under the paternalistic guise of protecting cotton growers from the unscrupulous Jula and at the same time preserving seed quality in the name of "development," CFDT succeeded in establishing itself as the sole buyer of Ivorian cotton. This monopsony allowed the cotton company to reimburse itself for the (inflated) costs of diffusing the new cotton package as well as to profit from controlling the ginning and sale of cotton. The emergence of the CFDT monopsony marked a new relationship between the state and private commercial interests. In contrast to the colonial period when French policy generally favored free trade and competition among merchant houses, the newly independent government sanctioned a monopolistic arrangement. As Bates persuasively shows, such monopsonies facilitate the transfer of capital from rural producers to other sectors of the economy or abroad, usually through the intermediary institution of government marketing boards.[3] Building upon late-colonial economic institutions (la Caisse de Soutien; la Caisse de Stabilisation), the Ivorian marketing board allied itself with private and semi-public agents like CFDT to intervene in new ways in the rural economy. This relationship between the Ivorian and

French states and private companies lasted some forty years after political independence in 1960. The CFDT system thus gestated and emerged out of the evolving policy discussions dating from the late colonial period on how to compete with the parallel cotton market and, at the same time, how to generate revenues to invest in projects for which the newly independent state needed financing. French government support for cotton development contained the political and economic objective of keeping its African colonies closely tied to the metropole through financing commodity development and price stabilization programs. French textile companies were particularly keen on rebuilding their war-torn industry, and they hoped to secure its essential raw material from France's African colonies.[4] Peasant farmers, on the other hand, equated cotton with forced cultivation and eventually had to be compelled by the national government to experiment with the new cotton package during the first decade of "independence."

Making cotton work: locally induced innovations

A second set of peasant farmer contributions to the cotton revolution center around locally induced innovations under changing social and economic conditions. Chapters 4, 5, and 6 demonstrated that the integration of cotton into savanna farming systems involved far-reaching social and cultural changes, as well as technological ones. While CFDT/CIDT-directed innovations have been biased towards increasing output per unit area, many locally induced innovations have been aimed at reducing labor bottlenecks, controlling household labor, and increasing incomes. The most important socio-cultural changes included flexible ways of interpreting culturally prescribed rest days, new forms of labor mobilization, and the greater importance of women's work in household fields. Tensions and conflicts over the direction and meaning of these changes have been a recurring theme in the cotton development story. The rise of conjugal households as the most common social unit of production, the *tyolobèlè* revolt in Katiali in the 1970s, and the shifting of women's personal field work days from the Senufo to the Roman calendar in the 1990s, are illustrative of the sometimes heated negotiations that have characterized the cotton revolution.

These fundamental changes in socio-cultural life were linked to transformations in the rural economy and a succession of technical innovations in local farming systems. Chapter 5 traced the changes in agricultural land use between the early 1960s and mid-1980s. The adoption of labor-saving technologies such as ox-plows and herbicides enabled many farmers to extend the area under both food crops and cotton. The expanded area significantly increased labor requirements for planting, weeding, and harvesting, which had the consequence of exacerbating labor constraints in the agricultural calendar. The dramatic decline in intercropping vividly captures the transformations taking place in local farming systems. The dream of colonial planners had come true. Increasingly large numbers of individual peasant farmers were devoting more resources to cotton production, and output levels soared.

Chapter 6 showed that the adverse economic conditions brought about by the elimination of the fertilizer and pesticide subsidies, falling producer prices, a more demanding cotton variety, and currency devaluation forced peasant farmers to rethink how to allocate their limited resources. They responded in both time-tested and innovative ways. The case of Katiali showed that the most common coping methods included crop diversification, the extensification of cotton, defaulting on loans, organized strikes, and the creation of new production and marketing cooperatives. As a result of these actions, CIDT was forced to modify its pricing policies, abandon its new glandless variety, and watch agricultural yields dramatically fall to levels more characteristic of the 1960s than of the 1980s. For the first time, cotton company officials in Bouaké and Paris began to speak with some respect for the power of cotton growers, whose collective actions made them take notice that business was not being conducted as usual. This recognition of peasant producers as important players in the cotton sector contrasts with their largely passive portrayal in the development literature. Nevertheless, it is the latter's image of peasant farming systems "waiting to be developed," of peasant farmers as subjects of development rather than as forces of innovation, that continues to dominate the cotton development discourse of the late 1990s.

The landscape of change

The thesis of this book is that the development of cotton in Côte d'Ivoire has been linked to the interplay of directed and induced technological and socio-cultural innovations that have developed in a dialectical and incremental manner since the early colonial period. The dynamics of agrarian change have been driven by negotiations among peasants and a multitude of external agents, as well as between them, over the nature and direction of these changes. These interactions were characterized by contradiction and conflict as much as by experimentation and innovation. From these interactions emerged new farming techniques and crop mixes, different forms of labor organization and conjugal relations, intra-household contestation over the allocation of farm work and income, and new modes of political mobilization. In short, the technological breakthrough proclaimed by CFDT as being the primary source of growth is viewed here as just one of many innovations that have been central to this story. The question remains, however, to what extent can the pattern of agricultural growth be considered revolutionary?

Linguists define the term "revolution" as "a sudden or momentous change in a situation."[5] The emphasis placed in this book on the incremental nature of agricultural change is contrary to the notion of a sudden change. But the change has been momentous, whether measured in terms of agricultural productivity, cropping patterns, agricultural technology, or in social and cultural terms. This said, these transformations have not been uniform throughout the savanna region. Considerable variation exists among as well as within regions in such areas as cotton yields and the use of ox-plows.[6] As Xavier Le Roy's comparative study of two communities in the Boundiali region suggests, the pattern of agricultural change has been uneven in both spatial and temporal terms.[7] Contrasting the case studies of Synofan and Karakpo, he shows that the two communities substantially differ in terms of area under cotton, the use of herbicides, and the degree of agricultural mechanization. Matty Demont and Philippe Jouve show a similar pattern of geographical variation in agricultural intensification in the Dikodougou region south of Korhogo.[8] Like Le Roy, they cite differences in demographic density, proximity to markets, and access to land as critical determinants in explaining this uneven development. However, their thesis that intensification follows a Boserupian path in which more intensive farming practices result from population pressure on natural resources does not adequately explain the dynamics of intensification in areas of low-population density, such as Katiali.

In addition, the case of Katiali shows important differences in the dynamics of intensification among social groups within communities. It demonstrates in particular that upper-income households were among the early innovators, that intra- and inter-generational conflicts over cash cropping were an important stimulus to intensification, and that women's labor contributions to household production (as opposed to their personal fields) were critical. Further evidence of the importance of social differentiation to this story is the uneven adoption of herbicides among households and the fissioning of cooperatives in order to segregate chronically indebted farmers. Local distinctions made between "good" and "bad" farmers suggest that not all farmers have been successful in making cotton work both agronomically and commercially. However, the longitudinal case study of Katiali shows that most households in the sample were farming with oxen, with the exception of two female-headed households which had to hire non-household ox-plow owners to work their fields.

Senior women residing in polygynous and upper-income households were more likely to cultivate cotton, use fertilizers and herbicides, and have their fields plowed by oxen, than their junior co-wives or women belonging to lower-income households. Unlike men, who obtained farm inputs from village cooperatives on credit, most women, especially Senufo women, obtained them in informal markets by providing in-kind services like weeding and harvesting the fields of the person providing the good or service.

To what extent have farmers benefited from this transformation of their farming systems, in which cotton has become the single most important crop? The two most telling graphs (Fig. 6.1 and Fig. 6.4) show that incomes in real terms steadily declined over the period 1970–94 and that cotton growers received a relatively small share of the market value of their crop. In comparison to cotton growers in Zimbabwe and India, West African farmers were at a distinct disadvantage. Profits were accruing to other agents in the cotton sector, notably to gin owners (CIDT), input suppliers, cotton traders (CFDT/COPACO), Ivorian textile firms, and the Côte d'Ivoire marketing board (Caistab).[9] The fiscal austerity measures pushed through during the 1990s (elimination of the pesticide subsidy, currency devaluation) created great hardship in the cotton growing areas. The surging number of indebted cotton growers and the drop in agricultural equipment sales testify to the recessionary impact of the World Bank's economic policies during this period. Rather than new metal roofs being placed on the modest homes of peasant farmers, a new layer of thatch was laid down.

The pattern of growth in Côte d'Ivoire cotton during the second half of the 1990s suggests that the period 1985–95 was not a denouement but an important transition in the unfolding of the cotton story. Figure 7.1 shows the upward trend in cotton production, area, and yields between 1994 and 1998. Seed cotton production rebounded from 217,000 metric tons in 1995/96 to 265,000 tons in 1996/97. The following year (1997/98) peasants produced 338,000 tons of seed cotton, establishing a new record.[10] The 1998/99 season set another record at 360,000 tons. When I visited Katiali in the summer of 1998, farmers were clearly pleased with the results of the 1997/98 season. Many spoke about how long it had been since they actually had money in their pockets. Some farmers were even putting new metal roofs on their homes. Cooperative record books indicated that the numbers of indebted growers had dropped to their lowest levels in more than a decade.

The increase in output was in part attributed to increased seed cotton prices, better performing varieties, well-distributed rainfall, improved yields, and a larger area in cotton.[11] Prices for first-grade seed cotton rose from 160 to 200 FCFA between the 1995/96 and 1997/98 seasons, although producers were still receiving less than half of the world market price.[12] Glandless cotton was eliminated in 1996/97 and replaced both by the more

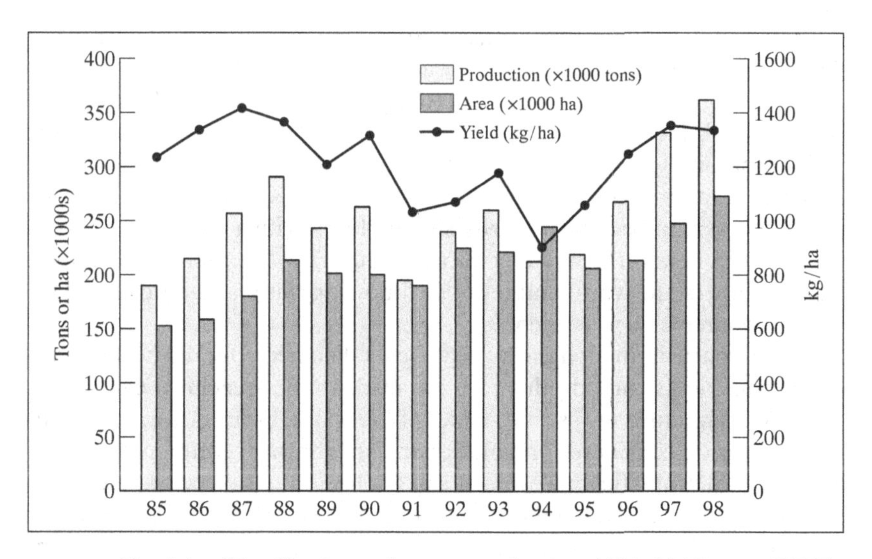

Fig. 7.1 Côte d'Ivoire seed cotton production, 1985–98 (*Source:* CIDT)

hardy and higher-yielding N'TA 88.6 variety imported from Mali and by the time-tested ISA variety. Although the yield data are problematic due to under-reporting of both area under cultivation and number of growers, the official figures show an approximate 300 kg/ha increase between 1995 (1,063) and 1998 (1,327).[13] Record production levels were also tied to a 25% increase in the area under cultivation (204,380 to 270,000 ha). The momentum had been regained. Exports were higher than ever in 1997/98, placing francophone African countries (as a group) as the number three exporter of cotton in the world that year.[14] Three more cotton gins were under construction in 1999/2000 in the newly privatized cotton economy, as the new stakeholders positioned themselves to profit from the cotton boom.[15]

The momentous nature of the cotton revolution was also evident in the growing power of tens of thousands of small farmers who, through producer organizations, continued to press for a higher share of the export price of cotton lint. During the first round of cotton seed price negotiations with private gin owners in 1998, producers refused to market their crop until a satisfactory agreement was reached that would automatically increase producer prices with any increase in world market prices.[16] Another strike was called by URECOS-CI in December 1999 to contest the move by new gin owners to lower producer prices as a result of higher than expected ginning costs and lower world market prices. The strike was called off a month later after the farmer organization agreed to the lower prices but only on condition that they be adjusted in March 2000, following a study of world market prices and actual ginning costs.[17] URECOS-CI, the strongest producer organization representing 85% of cotton producers and 90% of production,[18] also entered into a partnership with the French Louis Dreyfus Group to construct a new cotton gin in Korhogo. URECOS-CI members were to become majority shareholders, with a 60% capital investment in the new gin.[19] Whether rural producers will actually take home a larger share of the world market price following this major agro-industrial investment remains to be seen.[20]

In summary, in the long historical view taken here, impressive increases in cotton yields and production are evident. Far-reaching changes in farming techniques, crops, and modes of labor organization are also apparent and can be considered "revolutionary." In contrast, cotton growing has rarely led to dramatic improvements in the livelihoods of peasant farmers, who can count on one hand the number of years that they felt it was worth their effort. The question remains why farmers continue to invest their time and resources in this cash crop despite what one middle-income Senufo farmer, Bema Koné, declared to me in his two-room thatched roof home one July afternoon in 1995: "Cotton growing is zero, zero. After I sold my cotton this year, I didn't have enough money to buy clothes for my children, never mind me."[21]

Made of peasant cotton

One of the great paradoxes of the cotton revolution in Côte d'Ivoire is that it ever occurred. Contrary to the pessimistic assessments of scholars and aid donors writing on African agriculture during the 1980s and 1990s, the recent history of cotton in West Africa will strike many readers as surprising.[22] Some may remain skeptical about the "successful" nature of this development in the absence of a discussion on the environmental impact of cotton growing. Indeed, the notorious impact of cotton on soils, and the specter of pest resistance to insecticides in the West African savanna, demand careful research by agronomists and entomologists. Little serious study has been conducted in these basic areas despite there being good reasons to be concerned.[23] In the absence of such studies, the question still remains: to what extent is the case of West African cotton exceptional?

It is exceptional in the sense that influential agricultural policy analysts and development specialists like Robert Bates and Paul Richards have effectively argued that such an outcome was unlikely or exceptional at best. Although much of what they write on the political and human ecological bases of African agricultural policies and practices resonates deeply in the case of Côte d'Ivoire, their interpretations of the politics and innovative capacity of small-scale farmers do not adequately explain the kinds of agronomic and political developments documented in this study. Richards writes eloquently about the "delicately equilibrated labor economy of the typical 'catenary' farm," with an emphasis on the centrality of labor constraints in farmer decision-making on where and what crops to cultivate.[24] In his view, the introduction of labor-intensive crops like cotton poses serious constraints on peasant production and are doomed to fail. "There is no alternative to long-term participatory approaches and situation specific design," writes Richards, in which developers must come to recognize the logic of indigenous agricultural practices. Indeed, if CIDT had followed Richards's prescription, then perhaps the GL7 debacle would never have taken place. However, his agro-ecological conceptual framework, with its emphasis on equilibrium conditions, precludes the possibility of agricultural transformations that disrupt the "balance" between labor supplies and farming systems. In his view, farmers are unlikely to seriously consider a labor-intensive crop and radically adjust their farming system around it as the farmers of Katiali have with cotton. This study shows that farmers did not abandon their former modes of cultivation and social organization overnight to take up cotton. The new cotton package was progressively integrated into a rural economy which had already been greatly modified in the changing social, political, and economic conditions associated with the tumultuous times of the late nineteenth century and more than half a century of colonial rule.

While Richards views peasant experimental initiatives as focused on local fine-tuning to agro-ecological and labor conditions, the Katiali materials point to a more expansive notion of experimentation that includes the virtual replacement of one set of practices by another. An important component of this transformation has been the advantages offered by participating in the cotton development program. Although producer prices were low, they were consistently higher than those received by cotton growers in neighboring Mali and Burkina Faso. Other attractions included timely access to subsidized farm inputs (fertilizers, insecticides, herbicides), mechanization (ox-plows, winches for destumping fields), and credit. Non-cotton growers did not have equal access to these means of production and financing. Additional motivations included a guaranteed producer price for cotton and timely payments, a need for cash to pay off old debts as well as to satisfy new consumer needs, and the lack of an equally attractive cash crop.[25] In short, the opportunities and relative security offered by the cotton program largely outweighed the risks and uncertainty that Richards sees as the main reasons for peasant farmer resistance to far-reaching technological innovations. They also help to explain why Bema Koné continues to grow cotton and why his two junior wives wished to do so, even if informally.[26]

The cotton boom also appears to be an exceptional outcome to Robert Bates's assessment of the generally negative effects of public policies on agricultural production in Africa. However, Bates does make room for variation in agricultural policies on the basis of "meaningful party competition" and the stake that political leaders personally possess in the agricultural economy.[27] He specifically points to the case of Côte d'Ivoire, whose political elite had rural as opposed to urban ties and thus had a stake in maintaining a buoyant agricultural economy.[28] Yet Bates views peasant farmers as having little influence on agricultural policies. They are generally motivated by self-interest and are "'disengaged' from the national political arena," due to their sheer numbers and fear of repression."[29] In his view, peasants are more prone to free-riding than political mobilization.

The case of Côte d'Ivoire points to a more dynamic situation in which competing farmer organizations are effectively influencing agricultural policy. These new rural institutions emerged, in part, in the space opened up by political and economic reforms (e.g. multi-party politics and market liberalization) initiated by the Ivorian government to deal with its fiscal and political crises under pressure from the World Bank and IMF.[30] Structural adjustment was in this sense a double-edged sword. On the one hand, its privatization policies created space for farmer organizations at the price setting negotiating table and facilitated their entry into the industrial sphere of the cotton sector (ginning). However, the anticipated benefits of the currency devaluation did not trickle down to producers because of profit taking by CIDT and patrimonial politics manifested in cotton lint subsidies to the national textile industry. While a cross-section of farmers engaged in free-riding to weather the storm, they also used other tactics, such as market strikes and forceful negotiations on price policy committees, to press for their collective interests. It remains to be seen whether the farmer organizations will operate independently of formal party politics, whether they will become effective policy advocates for female as well as male farmers, and if they can acquire the financial resources and leadership that will enhance their legitimacy and political power.[31] The economic policy successes and growing strength of small-farmer organizations in Côte d'Ivoire suggests that peasant farmers will continue to be important players in the dynamics of agrarian change in West Africa.

Appendix 1: Côte d'Ivoire seed cotton production, 1912–1998

Year	Prod (×1000 t)	Area (× 1000 ha)	Yield (kg/ha)	Growers (1000s)	Principal Varieties
1912	0.12				
1913	0.376				
1914	0.227				
1915	1.213				
1916	1.591				
1917	0.564				
1918	0.434				
1919	NA				
1920	NA				
1921	0.051				
1922	NA				
1923	NA				
1924	1.26				
1925	2.54				
1926	4.23				
1927	5.15				
1928	6.01				
1929	5.74				
1930	6.82				
1931	4.48				
1932	3.57				Ishan, Karangi
1933	2.59				Ishan, Karangi
1934	4.05				Ishan, Karangi
1935	4.5				Ishan, Budi
1936	10.45				Ishan, Budi, Allen
1937	10.7				Ishan, Budi, Barbadense
1938	11.22				Ishan, Budi, Barbadense
1939	NA				Ishan, Budi, Barbadense
1940	1.69				Ishan, Budi, Barbadense
1941	3.7				Ishan, Budi, Barbadense
1942	6.79				Ishan, Budi, Barbadense
1943	3.71				Ishan, Budi, Barbadense
1944	1.53				Ishan, N'kourala, Togo Sea Is.
1945	0.7				Ishan, N'kourala, Barbadense

Year	Prod (×1000 t)	Area (× 1000 ha)	Yield (kg/ha)	Growers (1000s)	Principal Varieties
1946	0.5				Ishan, N'kourala, Barbadense
1947	0.06				Ishan, N'kourala, Barbadense
1948	1.7				Ishan, N'kourala, Barbadense
1949	NA				Ishan, N'kourala, Barbadense
1950	2.99				Ishan, N'kourala, Barbadense
1951	2.92				Ishan, N'kourala, Barbadense
1952	3.78				Ishan, N'kourala, Barbadense
1953	3.82				Ishan, N'kourala, Barbadense
1954	4.5				Ishan, Mono
1955	2.78				Ishan, Mono
1956	4.9				Ishan, Mono
1957	4.86				Mono
1958	NA				Mono
1959	NA				Mono
1960	0.07	0.14	504	NA	Mono, A151
1961	0.24	0.27	882	NA	A151, Mono
1962	0.77	1.3	599	NA	A151, Mono
1963	2.1	2.5	815	NA	A151, Mono
1964	5.5	6.4	863	NA	A151, Mono
1965	10	12	775	NA	A333-57, Mono
1966	25	24	926	NA	A333-57, Mono
1967	33	39	829	NA	A333-57
1968	41	48	867	62	A333-57
1969	32	33	969	44	A333-57
1970	29	36	817	47	Har444-2
1971	49	51	944	62	Har444-2
1972	53	57	935	67	Har444-2
1973	59	58	1005	68	Har444-2
1974	60	59	1023	69	Har444-2
1975	65	66	994	79	Har444-2
1976	75	65	1164	71	L231-24/69, L299-10/70
1977	103	88	1176	90	L231-24/69, L299-10/70
1978	115	107	1071	94	L299-10, T120-7
1979	143	123	1163	97	L299-10, T120-7
1980	137	126	1081	95	L299-10, T120-7
1981	135	125	1086	91	T120-7
1982	157	128	1223	91	T120-7
1983	142	136	1044	94	T120-7, ISA205
1984	212	146	1454	104	T120-7, ISA205
1985	189	153	1237	109	ISA205
1986	213	159	1335	113	ISA205
1987	256	180	1419	124	ISA205
1988	290	213	1360	145	ISA205
1989	242	202	1200	139	ISA205
1990	261	199	1315	124	ISA205, GL7
1991	194	191	1017	119	GL7, ISA205
1992	239	224	1066	133	GL7, ISA205

Year	Prod (×1000 t)	Area (× 1000 ha)	Yield (kg/ha)	Growers (1000s)	Principal Varieties
1993	258	219	1178	131	GL7, ISA205, 319
1994	210	242	865	151	GL7, ISA205, 319
1995	217	204	1063	138	GL7, ISA268, 319
1996	265	211	1259	138	NTA.88.6
1997	337	244	1380	153	ISA205, 268, 319, NTA 88.6
1998	361	271	1330	162	ISA205, 268, 319, NTA 88.6

Notes

1 INTRODUCTION

1 Groupe de Travail Coopération Française, *Le coton en Afrique de l'Ouest et du Centre: situation et perspectives* (Paris: Ministère de la Coopération Française, 1991). World Bank, *Cotton development programs in Burkina, Côte d'Ivoire, and Togo* (Washington, DC: The World Bank, 1988); Lele, U., N. van de Walle, and M. Gbetibouo, "Cotton in Africa: an analysis of differences in performance," *MADIA Discussion Paper 7*, (Washington, DC: The World Bank, 1989).

2 S. Berry, "The food crisis and agrarian change in Africa: a review essay," *African Studies Review* 27(1984), 59–112; World Bank, *Accelerated Development in Sub-Saharan Africa* (Washington, DC: The World Bank, 1984).

3 X. Le Roy, "Où la culture cotonnière rénove l'économie paysanne (Côte d'Ivoire)," in J.-P. Chauveau, M.-C. Cormier-Salem, and E. Mollard (eds.) *L'innovation en agriculture: questions de méthods et terrains d'observation* (Paris: IRD, 1999), p. 200. See also A. Sawadogo, *L'agriculture en Côte d'Ivoire* (Paris: Presses Universitaires de France, 1977), p. 135.

4 E. Roe, "Development narratives, or making the best of blueprint development," *World Development* 19(1991), pp. 287–300.

5 Lele, *et al.*, *Cotton*. The CFDT system is a variant of contract farming which Little and Watts define as "forms of vertical coordination between groups and buyers-processors that directly shape production decisions through contractually specifying market obligations (by volume, value, quality, and at times, advance price determination); provide specific inputs; and exercise some control at the point of production (i.e. a division of management functions between contractor and contractee)." P. Little and M. Watts, "Introduction," in P. D. Little and M. J. Watts (eds.), *Living under contract: contract farming and agrarian transformation in sub-Saharan Africa* (Madison: University of Wisconsin Press, 1994), p. 9.

6 Sawadogo, *L'agriculture*.

7 H. Bernstein, "African peasantries: a theoretical framework," *J. of Peasant Studies* 6 (1979), pp. 420–43.

8 A. Giddens, *Central problems in social theory: action, structure and contradiction in social analysis* (Berkeley: University of California Press, 1979), p. 69.

9 *Ibid.*, p. 70.

10 M. Watts, "Development I: power, knowledge, discursive practice," *Progress in Human Geography* 17 (1993), 257–72.

11 J. Ferguson, *The anti-politics machine: "development," depoliticization, and bureaucratic power in Lesotho* (Minneapolis: University of Minnesota Press, 1990), p. 71.
12 J. Ferguson, "The anti-politics machine: 'development' and bureaucratic power in Lesotho," *The Ecologist* 24 (1994), p. 177.
13 T. Mitchell, "The object of development," in J. Crush (ed.), *Power of Development* (London: Routledge, 1995), pp. 129–57.
14 J. Crush (ed.), *Power of Development* (London: Routledge, 1995), p. 10.
15 A. Escobar, *Encountering development: the making and unmaking of the Third World* (Princeton: Princeton University Press, 1995), p. 39.
16 A. Escobar, "Imagining a post-development era," in J. Crush (ed.), *Power of development* (London: Routledge, 1995), p. 213.
17 Escobar, *Encountering*, p. 47.
18 J. Crush (ed.), *Power*, p. 7.
19 S. Berry, *No condition is permanent: the social dynamics of agrarian change in sub-Saharan Africa* (Madison: University of Wisconsin Press, 1993), p. 48.
20 *Ibid.*, p. 189.
21 A. Isaacman and R. Roberts (eds.), *Cotton, colonialism, and social history in sub-Saharan Africa* (Portsmouth, NH: Heinemann, 1995).
22 R. Roberts, *Two worlds of cotton: colonialism and the regional economy in the French Soudan, 1800–1946* (Palo Alto: Stanford University Press, 1997); M. van Beusekom, "Contested development: African farmers, colonial officials, and agricultural practices at the Office du Niger, 1920–60," paper presented at the annual meeting of the African Studies Association, Columbus, Ohio, 13–16 Nov. 1997.
23 D. Maier, "Persistence of pre-colonial patterns of production: cotton in German Togoland," in A. Isaacman and R. Roberts (eds.), *Cotton, colonialism, and social history in sub-Saharan Africa* (Portsmouth, NH: Heinemann, 1995), pp. 71–95.
24 J. Hogendorn, "The cotton campaign in northern Nigeria, 1902–1914: an example of public/private planning failure in agriculture," in A. Isaacman and R. Roberts (eds.), *Cotton*, pp. 50–70.
25 O. Likaka, *Rural society and cotton in colonial Zaire* (Madison: University of Wisconsin Press, 1997).
26 J.-P. Chauveau, "L'étude des dynamiques agraires et la problématique de l'innovation," in J.-P. Chauveau, M.-C. Cormier-Salem, and E. Mollard (eds.), *L'innovation en agriculture: questions de méthods et terrains d'observation* (Paris: IRD, 1999), p. 21.
27 *Ibid.*, pp. 23–4.
28 S. D. Biggs and E. J. Clay, "Generation and diffusion of agricultural technology: a review of theories and experiences," *World Employment Programme Research Working Paper* 122 (Geneva: ILO, 1983), cited in P. Richards, *Indigenous agricultural revolution: ecology and food production in West Africa* (London and Boulder, CO: Hutchinson and Westview Press, 1985), p. 160. Richards also argues for an incremental approach to agricultural research and development.
29 D. Grigg, *The dynamics of agricultural change: the historical experience* (New York: Saint Martin's Press, 1982), pp. 165– 7.

30 S. Brush and W. Turner, *Comparative farming systems* (New York: The Guilford Press, 1987), p. 29.
31 Sawadogo, *L'agriculture*, pp. 68–9; Y-A Fauré, "Le complexe politico-économique," in Y-A Fauré and J-F Médard (eds.), *Etat et bourgeoisie en Côte d'Ivoire* (Paris: Karthala, 1982), p. 27.
32 R. Hecht, "The Ivory Coast miracle: what benefits for peasant producers?" *J. of Modern African Studies* 211 (1983), 25–83; J-M Gastellu, *Riches paysans de Côte-d'Ivoire* (Paris: L'Harmattan, 1989), pp. 101–3; J-M Gastellu and S. Yaffou Yapi, "Un mythe à décomposer: la 'bourgeoisie de planteurs'," in Y-A Fauré and J-F Médard (eds.), *Etat et bourgeoisie en Côte d'Ivoire* (Paris: Karthala, 1982), pp. 174–7; F. Ruf, *Booms et crises du cacao: les vertiges de l'or brun* (Paris: Karthala, 1995).
33 CIDT, *Rapport Annuel, Campagne 89/90* (Bouaké: CIDT, 1990), p. 12. The data for area per grower is for the period 1968–84. Prior to 1968, the number of cotton growers is unknown.
34 J. Amprou, L. Gbeli, and J. McIntire, "Le coton en Côte d'Ivoire: état des réformes, mimeograph, Banque Mondiale (3 juin 1998), p. 2.
35 The data on total production are still reliable given CIDT's market monopsony.
36 Grigg, *Dynamics,* pp. 174–6.
37 Brush and Turner (eds.), *Comparative.*
38 *Ibid.*, p. 40.
39 V. Ruttan, "Induced innovation and agricultural development," in G.K. Douglass (ed.), *Agricultural sustainability in a changing world order* (Boulder: Westview Press, 1984), p. 115.
40 *Ibid.*, pp. 119–23.
41 B. Koppel, "Why a reassessment?" in B. Koppel (ed.), *Induced innovation theory and international agricultural development: a reassessment* (Baltimore, Johns Hopkins University Press, 1995), pp. 3–21; R. Grabowski "Induced innovation: a critical perspective," in B. Koppel (ed.), *Induced innovation*, pp. 73–92; A. De Janvry, M. Fachamps, and E. Sadoulet, "Transaction costs, public choice, and induced technological innovations," in B. Koppel (ed.), *Induced innovation*, pp. 151–68.
42 V. Ruttan and Y. Hayami, "Induced innovation theory and agricultural development: a personal account," in B. Koppel (ed.), *Induced innovation*, p. 31.
43 Koppel, "Reassessment," p. 2.
44 J. Staatz and C. Eicher, "Agricultural development ideas in historical perspective," in *Agricultural Development in the Third World* (Baltimore, Johns Hopkins University Press, 1990).
45 L. Burmeister, "The South Korean green revolution: induced or directed innovation?" *Economic Development and Cultural Change* 35 (1987), pp. 766–90; L. Burmeister, "Induced innovation and agricultural research in South Korea," in B. Koppel (ed.), *Induced innovation*, pp. 39–55.
46 Burmeister, "Induced innovation," p. 44.
47 Burmeister, "South Korean," p. 785.
48 J-P Olivier de Sardan, *Anthropologie et développement: essai en socio-anthropologie du changement social* (Paris: Karthala, 1996), p. 91.
49 This interpretation of the social origins of CFDT/CIDT's cotton package differs from that of Xavier Le Roy, who views it as exogenous in origin. Le Roy, "Où la culture," p. 201.

50 Chauveau comes to a similar conclusion in his reflections on the relative importance of the notions of "spontaneous" and "provoked" innovations to agrarian studies. Chauveau, "L'étude," p. 21.

51 Jean-Pierre Olivier de Sardan's socio-anthropological view of innovation as a process of hybridization, reinterpretation, and reorganization is similar in this regard. However, his concept of innovation as an "original graft between different groups in an arena by way of go-betweens" ("*une greffe inédit, entre deux ensembles flous, dans une arène, via des passeurs*") does not sufficiently emphasize the power of local actors in shaping external innovations. Local actors are dynamic in how they "react" to exogenous innovations, but they do not appear to influence the nature of the introduced technique. Olivier de Sardan's interest is more in how peasants interpret, manipulate, and distort externally imposed innovations than in their role in shaping the very form and content of those innovations. Olivier de Sardan, *Anthropologie et développement*, pp. 77–96.

52 Richards, *Indigenous*, p. 141.

53 *Ibid.*, p. 118.

54 *Ibid.*, pp. 149–55. See Little, "The link", for an overview and critique of the participatory approach to rural development.

55 Richards, *Indigenous*, pp. 39–40.

56 *Ibid.*, pp. 66–8, 123.

57 To be fair, Richards does view a mix of "modern" and "traditional" farming techniques, such as applying commercial fertilizers on intercropped fields, as a particularly promising avenue for agricultural intensification. *Ibid.*, pp. 70–2.

58 *Ibid.*, pp. 17, 162.

59 Richards's oppositional categories of peasant and agricultural development agencies lead him to the paradoxical conclusion that "invention and ecological adaptation in African agriculture are their most vigorous where external agencies have interfered least." Richards, *Indigenous*, p. 14.

60 R. Bates, *Markets and states in tropical Africa: the political basis of agricultural policies* (Berkeley: University of California Press, 1981).

61 *Ibid.*, pp. 87–8.

62 Bates defines "free-riding" as a situation in which some producers devote time and resources to obtain higher prices which are subsequently enjoyed or undercut by other producers who did not engage in lobbying. This definition assumes the operation of a competitive market in which producers can sell their product at lower prices to increase the sales volume. This concept is of limited value in economies where prices are set by government agencies (e.g. marketing boards) or parastatal companies. The monopsonies typical of West Africa's cotton sector, and in many of the case studies cited by Bates, are not amenable to this type of free-riding.

63 *Ibid.*, pp. 82–7.

64 J. Scott, *Weapons of the weak: everyday forms of resistance* (New Haven: Yale University Press, 1985).

65 A. Isaacman, *Cotton is the mother of poverty: peasants, work and rural struggle in colonial Mozambique, 1938–61* (Portsmouth, NH: Heinemann; Cape Town: David Philip; London: James Currey, 1996).

66 Bates, *Markets*, p. 87.

67 Bates's discussion of the social, political, and institutional origins of the Mau

Mau rebellion in Kenya clearly shows that large numbers of rural households possess the capacity to organize and take collective action. R. Bates, *Beyond the miracle of the market: the political economy of agrarian development in Kenya* (Cambridge: Cambridge University Press, 1989), pp. 11–72.

68 Richards, *Indigenous*, p. 123.

69 To be fair, Bates believes that large numbers of small farmers could affect agricultural policies in the context of "meaningful party competition" in which political leaders promote the interests of peasants in order to gain their support in multi-party elections. This "important addendum" to Bates's otherwise pessimistic assessment of political action by small farmers is found in his essay, "The nature and origins of agricultural polices," in R. Bates, *Essays on the political economy of rural Africa* (Cambridge: Cambridge University Press, 1983), p. 131.

70 CIDT, *Annuaire signalétique: CIDT 1995–1996* (Bouaké: CIDT, 1996).

71 Groupe de Travail, *Coton*, p. 33.

72 J. M. Kowal and K. B. Adeoye, "An assessment of aridity and the severity of the 1972 drought in northern Nigeria and neighboring countries," *Savanna*, 2 (1973), pp. 146–7.

73 The maximum and minimum average monthly temperatures are 37 and 22 degrees centigrade respectively. The average monthly temperature equals 27 degrees centigrade. Insolation increases to 260 hours per month during the dry season (close to 9 hours per day) in comparison to 140 hours per month in July and August (less than 5 hours a day). Similarly, potential evapotranspiration reaches its peak levels during the dry season months (135–160 mm/month) then tapers off during the rainy season (100–112 mm/month) for a total loss of 1,525 mm/year. In light of these climatic conditions, the most favorable period for farming occurs from late May to mid-October. SEDES, *Région de Korhogo, Etude de Développement Socio-Economique,* Vol. 3 *Rapport Agricole* (Paris: SEDES, 1965), pp. 3, 22.

74 SEDES, *Région de Korhogo, Etude de Développement Socio-Economique,* Vol. 4, *Rapport Pedologique* (Paris: SEDES, 1965), pp. 14–18.

75 Between 1967 and 1974 ORSTOM carried out erosion studies in the Korhogo region and found that on bare soils on a slope of 4 degrees that the annual rate of erosion ranged between 3–9 tons/ha. GERDAT, *Intensification*, p. 10.

76 R. Launay, *Traders without trade* (Cambridge: Cambridge University Press, 1982).

77 T. Bassett, "Fulani herd movements," *Geographical Review* 76 (1986), 233–48.

78 A. Kientz, *Développement Agro-Pastoral et Lutte Anti-Tsé-Tsé, Côte d'Ivoire* (GTZ Report PN 87.2539.2–01.100; 1991), p. 23.

79 P. Bernardet, *Association agriculture-élevage en Afrique: les Peuls semi-transhumants de Côte d'Ivoire* (Paris: L'Harmattan, 1984).

80 T. Bassett, "The political ecology of peasant-herder conflicts in northern Ivory Coast," *Annals of the Association of American Geographers* 78 (1988), 453–72; T. Bassett, "Hired herders and herd management in Fulani pastoralism (Northern Côte d'Ivoire)," *Cahiers d'Etudes Africaines*, 34 (1994), 147–73.

81 Coulibaly, *Paysan*, 61–4.

82 J. Clifford, "Power and dialogue in ethnography: Marcel Griaule's initiation," in G. Stocking (ed.), *Observers observed: essays on ethnographic fieldwork*

(Madison: University of Wisconsin Press, 1983); P. Rabinow, *Reflections on Fieldwork in Morocco* (Berkeley: University of California Press, 1977); C. Keyes Aderoike and J. Vansina (eds.) *In pursuit of history: fieldwork in Africa* (Portsmouth, NH: Heinemann, 1995).

83 T. Bassett, "Food, peasantry and the state in the northern Ivory Coast, 1898–1982," Ph.D. thesis, University of California, Berkeley (1984).

84 T. Bassett, "Development theory and reality: the World Bank in Northern Côte d'Ivoire," *Review of African Political Economy*, 41 (1988), 45–59.

85 World Bank, *Accelerated*; K. Cleaver and G. Schreiber, *Reversing the spiral: the population, agriculture, and environment nexus in sub-Saharan Africa* (Washington, DC: The World Bank, 1994).

86 Bernstein, "African"; B. Dinham and C. Hines, *Agribusiness in Africa* (London: Earth Resources Research, 1983).

87 P. David, "La carte postale ivoirienne de 1900 à 1960: un bilan iconographique et culturel provisoire," *Notes Africaines* 174 (1982), pp. 29–39.

2 THE COLLISION OF EMPIRES, 1880–1911

1 T. Bassett and P. Porter, "'From the best authorities': the mountains of Kong in the cartography of West Africa," *J. of African History,* 32 (1991), 367–413.

2 R. Caillié, *Journal d'un voyage à Temboctou et à Jenné dans l'Afrique centrale* (Paris: l'Imprimerie Royale 1830).

3 It is Binger who is credited with "discovering" the Senufo ("Sien-ré"), although Louis Tautain had identified them earlier as a distinct ethnic group whom he called "Sénéfo". "Rapport sur le Concours au Prix Annuel fait à la Société de Géographie dans sa séance générale du 25 avril 1890," *Bulletin de la Société de Géographie* (Paris), Vol. 11 (1890), 151. L. Tautain, "Le Dioula-dougou et le Sénéfo," *Revue d'Ethnographie* VI (1887), 395, cited in R. Launay, "Stereotypic vision: the 'moral character' of the Senufo in colonial and postcolonial discourse," *Cahiers d'Etudes Africaines* 49 (1999), 272.

4 Caillié, *Journal,* p. 14.

5 *Ibid.*, pp. 4, 6, 18, 50, 57–8, 63.

6 *Ibid.*, p. 69.

7 *Ibid.*, pp. 87–8.

8 *Ibid.*, p. 99.

9 *Ibid.*, p. 86.

10 L. G. Binger, *Du Niger au Golfe du Guinée* (Paris: Hachette, 1892), Vol. 1, pp. 178–80. As a result of the Tingréla incident, many Parisians believed that Binger was dead.

11 Binger, Le Capitaine, "Du Niger au Golfe de Guinée," *Le Tour du Monde* LXI (1891), I, p. 68.

12 Binger, *Niger*, Vol. 1: p. 250.

13 *Ibid.*, p. 264.

14 *Ibid.*, p. 269.

15 C. L. Brubaker, F. M. Bourland, and J. F. Wendel, "The origin and domestication of cotton," in C.W. Smith and J. T. Cothren (eds.), *Cotton: origin, history, technology, and production* (New York: John Wiley and Sons, 1999), pp. 3–31.

16 Hutchinson believed that *Gossypium herbaceum* subsp. *africanum* was the

living ancestor of modern cotton varieties. Jonathan Wendel believes that *Gossypium herbaceum* subsp. *africanum* is a modern descendant of the wild A-genome ancestor. (J. B. Hutchinson, R. A. Silow, and S. G. Stephens, *The Evolution of Gossypium* [London: Oxford University Press, 1947], pp. 34–6; personal communication, Jonathan F. Wendel, 24 May 1999). Also see J. F. Wendel, "Cotton: Gossypium (Malvaceae)," in J. Smartt and N. W. Simmonds (eds.), *Evolution of crop plants* (London: Longman, 1995), pp. 358–66.

17 A. N. Prentice, *Cotton, with special reference to Africa* (London: Longman, 1972), p. 66.

18 New World cottons are tetraploids (fifty-two chromosomes) derived from Old World A-diploids (twenty-six chromosomes) that combined with New World D-diploids (twenty-six chromosomes). J. M. Munro, *Cotton* (New York: Longman, 1987), p. 32. There is much speculation on how *africanum* arrived in the New World. Munro summarizes the major hypotheses (Munro, *Cotton*, pp. 2–4).

19 In fact, little distinction was made between the "scientific" and "economic" objectives of this fifteen-month long expedition. Gatelet and Chevalier each emphasized both dimensions of the mission in their accounts. Among the mission's members were an engineer (Roné) who came to study the hydroelectric potential of certain stretches of the Niger River; the botantists Hamel and Chevalier, who were to study the economic importance of wild vine rubber; and two representatives of the French textile industry (Jacquery and Fossat), whose objective was to collect cotton specimens and conduct field trials. Chevalier was also involved with identifying cotton species, their geographic distribution, and the selection of varieties. A. Chevalier, "L'avenir de la culture du cotonnier au Soudan Français," *Bulletin de la Société Nationale d'Acclimatation de France*, (1901), pp. 225–44; Lt. Gatelet, *Histoire de la Conquête du Soudan Français (1878–1899)* (Paris: Berger-Levrault and Compagnie, 1901).

20 Chevalier, "L'avenir," pp. 231–34; A. Chevalier, "La question de la culture des cotonniers en Afrique tropicale," *Comptes rendus des séances de l'Academie des Sciences* (1904), p. 2.

21 Hutchinson, *et. al.*, *Evolution,* p. 40.

22 S. G. Stephens, "The origins of Sea Island cotton," *Agricultural History* 50 (1976), pp. 391–9. Annual forms were selected to produce bolls before the first killing frost. Perennial forms had a longer growing cycle, which meant that plants died off before they could bear fruit (Hutchinson, *et al.*, p. 103).

23 J. Hutchinson, *The application of genetics to cotton improvement* (Cambridge: Cambridge University Press, 1959), p. 32.

24 The *punctatum* variety has a high proportion of tufted seeds in which the fuzz layer is absent. This layer impedes the removal of seeds during hand-ginning. One indication of its popularity in the New World is that 80% of the seeds recovered in an archaeological site in Mexico (AD 700–1300) were tufted. Brubaker, *et al.*, "The origin," p. 25.

25 Chevalier, "L'avenir," p. 233; J. Lebeuf, "Le coton en Côte d'Ivoire et en Haute Volta" in E. Guernier (ed), *L'Encyclopédie Coloniale et Maritime*, Vol. 2, *Afrique Occidentale Française* (Paris: Encyclopédie Coloniale et Maritime, 1949); Prentice, *Cotton,* p. 4.

26 Prentice, *Cotton*, p. 4. Jonathan Wendel believes that the New World cottons are *more* than fifty percent African since the nucleus in which the two genomes reside is derived from an A-genome cellular matrix or cytoplasm. J. Wendel, personal communication, 24 May, 1999.
27 Hutchinson, *Application*, p. 35.
28 ANSOM Aff Pol 3047 bis, d. 39, "La question du coton dans la N'zi Comoé de 1914 à 1917."
29 William Bartram points to this labor-saving aspect of perennial cotton in his observations on cotton growing in the West Indies. W. Bartram, *The Travels of William Bartram*, ed. M. Van Doren (New York: Facsimile Library, 1940), pp. 75–9, cited in Stephens, "Sea Island Cotton," p. 396.
30 Roberts, *Two worlds*, pp. 210–14.
31 A. S. Kayna-Forstner, *Conquest of the western Soudan: a study in French Military Imperialism* (Cambridge: Cambridge University Press, 1969).
32 Y. Person, *Samori: une révolution Dyula*, Mémoires de l'Institut Fondamental d'Afrique Noire, 80, Vol. 3 (Dakar: IFAN, 1968), p. 1610.
33 Person, *Samori,* Vol. 2, p. 876; ANRS 1 G 146, Mission du Capt. Quiquandon dans le Kénédougou, 1890–1891, p. 12.
34 C. Rondeau, C., "La société Sénoufo du sud Mali (1870–1950): de la 'tradition' à la dépendance," Thèse de doctorat de 3eme cycle, Université de Paris (1980), pp. 289–92
35 Binger, *Niger,* pp. 238, 262.
36 ANSOM AF 1142, Service Géographique des Colonies, Mission Marchand, Le Transnigérien, Le Bandama et le Bagoé, (la 1ère Partie le Bani-Niger a été comprise dans la Carte Binger), Carte Levée et Dresée de 1892–1895 par le Capitaine Marchand, 1/500.000.
37 ANSOM Missions 8, Le Capt. Marchand de l'Infanterie de Marine, chargé de mission en Afrique, A Monsieur le Ministre des Colonies à Paris, Thiassalé, le 20 décembre 1894.
38 *Ibid.*
39 J. Méniaud, *Les Pionniers du Soudan* (Paris: Société des Publications Modernes, 1931), Vol. 2, pp. 486–90. See A. S. Kayna-Forstner's classic *The Conquest of the Western Sudan,* for details about French military strategy and presence in the "boucle de Niger" at this time.
40 Sinali Coulibaly, "Etat, société et développement: le cas Sénoufo dans le nord Ivoirien," Thèse de doctorat d'état, Université de Paris X – Nanterre (1990), p. 126.
41 Coulibaly, *Etat*, p. 126.
42 Coulibaly, *Etat*, pp. 126–9.
43 Person, *Samori*, Vol. 3, p. 1577.
44 Person, *Samori*, Vol. 3, p. 1578.
45 Person, *Samori*, Vol. 3, pp. 1739–40.
46 Kayna-Forstner, *Conquest*, p. 252.
47 ANSOM, S.G. Soudan, dossier 9, letter from the Governor General of AOF to the Minister of the the Colonies, 31 March 1898.
48 Person, *Samroi,* Vol. 3, pp. 1758–9.
49 Person, *Samori*, Vol. 3, p. 1788, FN 83.

50 J. Méniaud, *Pionniers*; Kayna-Forstner, *Conquest*; Person, *Samori*, Vol. 3.
51 ANSOM, S.G. Sudan, dossier 9, Rapport politique au mois de mai, 8 juin 1898.
52 ANRCI, Monographes, "Essai de Monographie du Cercle de Korhogo,"par Maurice Delafosse, 1905, p. 11.
53 ANSOM, Soudan 4, Letter from M. Ballay, Inspector of the Colonies to the Minister of the Colonies, 13 July 1898.
54 ANRCI, "Essai," pp. 11–12.
55 ANSOM, S.G. Sudan, dossier 9, letter from the Lt. Gov. of the French Sudan to the Min. of the Colonies, 28 April 1899, a/s envois de la carte politique du Soudan Français.
56 See Claude Meillassoux, *Anthropologie économique des Gouros de Côte d'Ivoire* (Paris: Mouton, 1964), pp. 292–5; Timothy Weiskel, *French colonial rule and the Baoule peoples: resistance and collaboration, 1889–1911* (Oxford: Clarendon Press, 1980).
57 ANRM 1N-174, Operations militaires, Rapports Côte d'Ivoire, 1890–1899.
58 For example, the quarter of Sedionkaha was founded by refugees from the former village of Segbaran. A number of blacksmiths from Diemé resettled in the quarter of Fononkaha, while the new quarters of Zambédala and Gborplédala were founded by inhabitants of the former villages of Katialiplé and Selio respectively.
59 ANRCI 2704, Rapport politique, Année 1958, par D. Pinelli.
60 ANRCI 426, Rapport des tournées effectuées du 25 avril au 30 mai 1942 dans le Cercle de Korhogo par l'Administrateur du Cercle 3 Juin 1942, p. 8.
61 Coulibaly, *Etat,* p. 178.
62 *Ibid.*
63 Circle Commanders and Subdivision Chiefs also maintained a personal encampment in each canton for their rare trips into the bush. Villagers were obligated to maintain these quarters or face punishment.
64 ANRCI 3333, "Rapport d'inspection administrative du Poste de Korhogo," 1933.
65 The term is derived from the Arabic word *gaum* meaning "troop." In French colonial history, a *Goum* referred to a military contingent furnished by a local group to the French army in Algeria. "Goumier" refers to a "cavalier d'un goum" – a trooper or cavalryman. *Petit Robert Dictionnaire* (Paris, 1981), p. 877.

3 THE UNCAPTURED CORVÉE 1912–1946

1 Anon., "L'industrie cotonnière française et les pays d'outre-mer," *Revue Internationale des Produits Coloniaux* 26 (1951), pp. 80–1, 83.
2 Launay, *Traders,* p. 83.
3 Peasant resistance to forced cultivation did play a role in frustrating the sometimes grandiose schemes of colonial planners. The most common forms of this resistance – foot-dragging and field neglect – were relatively safe ways for peasants to defend their interests against the state. Although the collaboration of chiefs and the brutality of circle guards enforcing cotton policies limited the scope and effectiveness of these acts of "everyday resistance," there is evidence

of communities taken aggressive stances towards the colonial state which involved suppression. Such overt and especially covert struggles against the cotton corvée are a salient thread running through this story.

4 Roberts, *Two Worlds*, p. 287.

5 Letter from Min. of Colonies to the Gov. Gen. of AOF on the subject of the French Soudan, instructions for the Lt. Gov. de Trentinian, dated 10 November 1898, ANSOM, S.G. Soudan I, dossier 9.

6 A. Chevalier, "La question," pp. 1–3.

7 "Rapport de Tournée, 1ère partie, Korhogo à Mankono, 1–13 mars 1907" Musée de l'Homme, Bibliothèque, MS 241, Haute Côte d'Ivoire, Mélanges Manuscrits.

8 ANRCI 812, Association Cotonnière Coloniale, "Status". The Association Cotonnière Coloniale was founded in 1902 by French textile concerns (spinners, weavers) to promote the development of cotton in France's overseas colonies, principally in West Africa.

9 ANSOM Aff Pol 3047 bis, Situation de la Culture du Cotonnière et de la Production du Coton au 31 décembre 1917 (Bingerville 1918), p. i. Emphasis added.

10 Isaacman and Roberts, *Cotton*, p. 18.

11 Ansom Aff Pol 3047 bis, "Situation," p. 26.

12 ANRCI 897b, Serv. de Textiles de Bouaké, Rapport du ferme cotonnière de Bouaké, 1932.

13 The techniques described by Delafosse in 1905 are still common today. See M. Delafosse, "Le Peuple Siéna ou Sénoufo," *Revue des études ethnographiques et sociologiques*, (1908), 260.

14 Launay, *Traders*, p. 83.

15 ANRCI 1 QQ 81, "Situation agricole de Korhogo, 1909," p. 1.

16 ANRCI 1 RR 63d, "Correspondances relatives à la campagne cotonnière et aux maladies du cotonnière."

17 ANRCI MiFm 5G 67, letter of Lt. Gov. Angoulvant on the political, economic and administrative situation of Côte d'Ivoire 26 November 1908.

18 ANSOM AP 3047 bis, d. 39, "La question du coton dans la N'zi Comoé de 1914–à 1917," p. 16.

19 Interview with Bêh Tuo, Katiali, 3 July 1981.

20 A. Isaacman, *Cotton is the mother of poverty: peasants, work, and rural struggle in colonial Mozambique, 1938–1961* (Portsmouth, NH: Heinemann, 1986); Likaka, *Rural society*, pp. 46–56.

21 ANRCI 1 QQ 81b, Cercle de Korhogo, Rapport économique, 23 April 1912.

22 ANRS 1 R 108a, "Circulaire sur l'intensification de la production agricole et pastorale en Côte d'Ivoire," 4 April 1931.

23 In the mid-1920s, Senufo peasants in the Sikasso area of Mali were required to grow ten kilograms per taxpayer. Rondeau, *Senufo*, p. 411.

24 ANRCI 1 RR 63c, Correspondances relatives au plan de campagne de culture cotonnière dans les Cercles N'zi-Comoé, Baoulé, Ouorodougou, Tagouanas et Kong, letter from Lt. Gov. of C I. to Admin. of Circles of 31 January 1918; letter from the Head of the Agric. Serv. to Lt. Gov. of C. I. of 16 March 1917.

25 ANRCI 1 RR 63c, letter from Lt. Gov. of C. I. to Admin. of Circles of January

31, 1918; letter from the Head of the Agric. Serv. to Lt. Gov. of C. I. of 16 March 1917.

26 Interview with Gniofolotien Silué, Katiali, 3 March 1992.

27 ANRCI 1 QQ 81d, Cercle de Kong, Poste de Korhogo, Note succinte sur la situation agricole, 6 December 1916.

28 T. Ouattara, "Les Tiembara de Korhogo, des origines à Péléforo Gbon Coulibaly (1962)" (Thèse de Doctorat de 3eme cycle d'Histoire, Université de Paris, 1977).

29 ANRCI 1 EE 79(5)b, Cercle de Kong, Poste de Korhogo, Rapport Mensuel, November 1915.

30 *Ibid.*

31 ANRCI 1 RR 14, Correspondance au sujet de l'intensification de la production agricole pour ravitailler la metropole – 1916–1918; letter from Gov. General, 7 June 1917.

32 ANRM lR-123, Note sur la culture de coton en Côte d Ivoire, 1918.

33 *Ibid.*

34 ANSOM, Aff Pol 3047 bis, H. Leroide, *Situation de la culture du cotonnier et de la production du coton au 31 décembre 1917* (Bingerville: Colonie de la Côte d'Ivoire, 1918), p. 29.

35 *Ibid.*

36 Leroide, *Situation,* p. 29. Also, see C. Monnoyer, "Le problème de la Culture du Cotonnier en Afrique occidentale française," *Revue de Botanique Appliquée et d'Agriculure Coloniale* 2 (1922), p. 613.

37 ANRCI 1 RR 63c, letter from Lt. Gov. of C. I. to Admin. of Circles of 31 January 1918; letter from the Head of the Agric. Serv. to Lt. Gov. of C. I. of 16 March 1917.

38 ANRCI 1 RR 63a, Situation de la culture de coton en Côte d'Ivoire, Rapport sur la culture de cotonnière: Plan de campagne, 1913.

39 ANRCI 1 RR 63m, Rapport de tournée par L. Leraide, Chef du Service de l'Agriculture, 15 March 1915.

40 ANRS 1 R 48a, La culture de coton dans les différentes colonies de l'AOF, 1903–30.

41 ANRS 1 R 48, Vers 158, Enquête sur la culture du coton, 27 August 1918.

42 ANRS 1 R 48, Vers. 158, Letter from Gov. Gen. to the Gov. of CI, 5 Sept 1918.

43 ANRS 1 R 48, Vers. 158, Letter from Gov. Gen. Angoulvant to Gov. Antonetti, 31 October 1918.

44 ANRCI 1 RR 63, Rapport de Mon. Waddington sur l'état actuel et de l'avenir de la culture cotonnière en Afrique Occidentale, 3 February 1924.

45 ANRCI 1 RR 67a, Correspondances relatives à l'envoi des semences, à l'achat des semences et à l'intensification de la culture du coton dans le cercle de N'zi-Comoé, 14 February 1925.

46 ANRCI 1 RR 67b, Circulaires relatives au développement de la culture du coton dans le cercle de N'zi-Comoé, 7 May 1925.

47 ANRS, 1 R 48a, 'La culture de coton dans les différentes colonies de l'AOF, 1903–30.

48 *Ibid.*

49 ANRCI 1 RR 63, Rapport de Mon. Waddington sur l'état actuel et de l'avenir de la culture cotonnière en Afrique Occidentale, 3 February 1924.

50 ANRCI, 1 RR 64b, Campagne cotonnière en Côte d'Ivoire, 1925–7.
51 *Ibid.*
52 Roberts, *Two Worlds*, pp. 192–222.
53 ANRCI 598, Letter from the Gov. of Côte d'Ivoire to the Gov. Gen. of AOF, a/s plan de production cotonnière, 24 Sept 1938; ANRCI 813, Cercle de Kong, Campagne Cotonnière 1935–6.
54 ANRCI 3442, Campagne cotonnière 1928.
55 ANRCI, 1 RR 67a, Correspondances relatives à l'envoi des semences, à l'achat des semences et à l'intensification de la culture du coton dans le cercle de N'zi-Comoé, 14 February 1925; ANRCI 1 RR 64b, Campagne cotonnière en Côte d'Ivoire, 1925–7.
56 A small number of women were also recruited to serve as cooks for work gangs. According to one informant in Katiali, this was the worst forced labor experience endured by women. Interview with Namwa Koné, Katiali, 12 March 1992.
57 Richard Roberts provides a detailed description of the organization and coercive aspects of these "fairs" in the French Soudan around the same period. Roberts, *Two Worlds*, pp. 182–8.
58 A. Londres, *Terre d'Ebène* (Paris: Arléa, 1998 (1928)), 135–7. Londres's book caused such a stir that the Governor-General of West Africa organized a press tour to address public concerns about the scandalous conditions brought to light by Londres. Photos taken Albert Londres during his 1928 Africa tour were discovered by Didier Folléas in a Casablanca flea market and subsequently published by Arléa. See D. Folléas, *Putain d'Afrique: Albert Londres en Terre d'Ebène* (Paris: Arléa, 1998). Plate 8 is reprinted from Folléas's book (p. 13) with the kind permission of Arléa.
59 ANRCI 2990, Inspection des Affaires Administratives, Rapport à Mon. le Gov. de la CI a/s des usages commerciaux particuliers au Cercle de Korhogo et difficultés qui sont produits entre les maisons de commerces et l'Administrateur M. Lalande, 26 January 1931.
60 ANRCI 1115, Service des Textiles, 1928–37.
61 ANRCI 1 RR 65, Correspondance relative aux champs d'essais de coton dans les cercles – 1927–28; telegramme officiel de l 'Inspecteur, Serv. des Textile, 30 April 1928.
62 C. Aubertin, *Histoire et création d'une région "sous-developpée" (le Nord Ivoirien)*, Abidjan: ORSTOM, 1980), pp. 18–22.
63 ANSOM AP 3066, Inspection des Affaires Administratives, Kair Report, 4 March 1931.
64 ANRCI 840a, Service des Textiles, Correspondance avec les cercles du nord; letter from Lt. Gov. of CI to Admin. of Korhogo Circle on the subject of peanut cultivation, 15 April 1929.
65 ANRCI 2896, L'affaire Msgr. Diss, relative à la pretendu famine dans le Cercle de Korhogo, 12 September 1931.
66 ANRCI AP 3066, 'Kair Report,' p. 28–33.
67 *Ibid.*, p. 39.
68 *Ibid.*, pp. 40–1.
69 *Ibid.*, pp. 49–51.
70 *Ibid.*; ANRCI 2990, Inspection des Affaires Administratives, Rapport à Mon.

le Gov. de la C. I. a/s des usages commerciaux particuliers au Cercle de Korhogo et difficultés qui sont produits entre les maisons de commerces et l'Administrateur M. Lalande, 26 January 1931.

71 ANRS 1 R 108c, Cercle de Kong, Rapport sur l'exécution de la première tranche du programme agricole quinquennal, 5 July 1932.

72 ANRCI 1 RR 67c, Correspondance relative à la vente, à la production du coton et aux matériels textiles, 1912–31.

73 ANRS 1 R 108d, letter from the Gov. of CI to Gov. Gen. of AOF, 11 August 1932.

74 ANRCI 840b, Rapport de l'Ingénieur des Etages sur son tournée effectuée dans la région septentrionale de la Côte d'Ivoire, 28 October 1935.

75 ANRCI 861, Cercle de Kong, Situation agricole, 1935.

76 M. Perron, "Situation économique et agricole du pays Sénoufo: notre rôle," *Bulletin de l'agence générale des colonies* (1933), p. 68.

77 Aubertin, *Histoire,* p. 14.

78 The Budi variety was a cross between *Gossypium hirsutum* and *Gossypium arboreum* locally known as Garrah Hills and Karangani. It was introduced from India and developed at the M'Pesoba experiment station in the French Soudan by the colonial agronomist Budichowsky, who worked for the Office du Niger in the Soudan. Budi was cultivated in the area south of Bamako and especially in the Bougouni Cercle. R. Clerin, "La production cotonnière au Soudan Français" in E. Guernier (ed.), *L'Encyclopedie Coloniale et Maritime*, Vol. 2, *Afrique Occidentale Française* (Paris: Encyclopédie Coloniale et Maritime, 1949). See also Roberts, *Two Worlds*, p. 253.

79 ANRCI 840b, Rapport de l'Ingénieur des Etages sur son tournée effectuée dans la région septentrionale de la Côte d'Ivoire, 28 October 1935.

80 ANRCI 1 RR 67d, Rapport des tournées effectuées par l'Ingénieur des Etages dans la zone de culture d'Ishan en vue de l'étude des questions cotonnière et principalement des maladies, 1936.

81 ANRS 1 R 108e, Letter from Gov. Gen. de Coppet to the governors of the French Soudan, Sénégal, Côte d'Ivoire, and Dahomey on the subject of bush cotton, 19 February 1938.

82 ANRS 1 R 108f, Programme de développement de la culture cotonnière en AOF, 20 August 1938.

83 Frederick Cooper, *Decolonization and African society: the labor question in French and British Africa* (Cambridge: Cambridge University Press, 1996), pp. 85–8.

84 ACCCI, Dossier 24, Ets. R. Gonfreville, Bouaké, Côte d'Ivoire, 'Culture de Coton en Côte d'Ivoire', October 1951.

85 ANRS 1 R125, Recherche et Sélection d'une variété de cotonnier adaptée aux conditions climatologiques et agrologiques locales: Travaux effectués à la Ferme Cotonnière de Ferkésédougou de 1926–38, Rapport par J. Lebeuf, 8 June 1941.

86 ACCCI, Dossier 24, Ets. R. Gonfreville, Bouaké, Côte d'Ivoire, Culture de Coton en Côte d'Ivoire, October 1951.

87 ANRS 1 R 208, Vers. 158, Organisation de la Production Cotonnière en AOF (1942). For a discussion of Vichyism and Free France in West Africa during the Second World War, see Jean Suret-Canale, *French Colonialism in Tropical Africa, 1900–1945* (New York: Pica Press, 1971), pp. 462–77.

88 Interview with Gniofolotien Silué, Katiali, 3 March 1992.
89 For a detailed discussion of colonial labor policies in French West Africa during this period, (1936–46), particularly the debates around forced versus "free" labor, see Cooper, *Decolonization*, pp. 77–88, 141–70.
90 ANRCI 3039, title page missing.
91 ANRCI 3039, Letter from the Head of the Boundiali Subdivision to the Commander of the Korhogo Circle, 6 August 1943.
92 ANRCI 3039, Letter No. 2084, from the Commander of the Korhogo Circle to the Gov. of Côte d'Ivoire, 14 August 1943.
93 ANSOM 1 Mi 2696, 2 G 43 (99), Rapport sur la situation politique d'ensemble pendant le mois de mai 1943.
94 Up until 1929, Katiali was the seat of the Canton of Katiali.
95 ANRCI 426, Rapport des Tournées effectuées du 25 avril au 30 mai 1942 dans le Cercle de Korhogo par l'Administrateur du Cercle, 3 Juin 1942.
96 ACCCI, Dossier 24, L. Chabrand, Note sur le coton en Côte d'Ivoire, Sept. 1944.
97 *Ibid.*
98 The resurgence of forced labor in Côte d'Ivoire during the reign of Pierre Boisson, the Vichy Governor-General of French West Africa, contrasts with Patrick Manning's view that "taxes were low and there were few demands for forced labor" during the early war years. See Patrick Manning, *Francophone Sub-Saharan Africa, 1880–1985* (Cambridge: Cambridge University Press, 1988), pp. 138–9. Frederick Cooper demonstrates that in contrast to the ambiguous discourse of the Popular Front administration on the issue of forced labor, the Vichy government openly discussed it and saw little alternative to it. Cooper, *Decolonization*, pp. 145, 153.
99 Latrille became governor of Côte d'Ivoire in August 1943. At the Brazzaville conference of colonial governors, convened in January of 1944 by de Gaulle and Félix Eboué, the governor-general of French Equatorial Africa, recommendations were made to end forced labor and to extend trade union rights in the colonies.
100 ACCCI, Dossier 24, Ets. R. Gonfreville, Bouaké, Côte d'Ivoire, Culture de Coton en Côte d'Ivoire, October 1951.
101 ANSOM 14 Mi 1848, 2 G 43 (51), Côte d'Ivoire, Service de l'Agriculture, "Rapport Annuel 1943".
102 ANSOM 14 Mi 1856, 2 G 44 (50), Côte d'Ivoire, Service de l'Agriculture, "Rapport Annuel 1944".
103 ANSOM 14 Mi 1878, 2 G 46, Colonie de la Côte d'Ivoire, Service de l'Agriculture, "Rapport Annuel 1946."
104 Roberts, *Two Worlds*, p. 30.

4 REPACKAGING COTTON, 1947–1963

1 ANRCI 840, Service de Textiles, Correspondance avec les cercles du nord; rapport de l'Ingénieur des Etages sur sa tournée effectuée dans la région sepentrionale de la Côte d' Ivoire, 28 October 1935.
2 ANS-AOF 2 G 47–68, Côte d'Ivoire, Service de l'Agriculture, *Rapport Annuel – 1947*, p. 65.
3 *Ibid.*

4 *Ibid.*
5 ANRCI 598, Letter from Gov. Pechoux to the High Commissioner of French West Africa, 5 February 1949.
6 ACCCI, Dossier 24, Culture du Coton en Côte d'Ivoire, Note 10.
7 ANRCI 598, Letter from the High Commissioner to the Minister of Overseas France, a/s Qualité et prix coton Cotivoire, 6 January 1949.
8 ACCCI, Dossier 24, Culture du Coton en Côte d'Ivoire, October 1951.
9 ANRCI 598, Letter from Gov. Pechoux to the High Commissioner of French West Africa, 5 February 1949.
10 ACCCI, Dossier 24, Culture du Coton en Côte d'Ivoire, October 1951.
11 ACCCI 324, CFDT, Conditions du développement de la production cotonnière en Côte d'Ivoire, Mai 1952.
12 ACCCI 24, Procès verbal de la Réunion de la Conférence Annuelle du Coton," 13 October 1951, p. 3.
13 ANRS 1 R 208, Ver. 158, Territoire de la Côte d'Ivoire, Circulaire du Service de l'Agriculture a/s culture cotonnière, February 1951, p. 1.
14 The Gonfreville textile mill was founded in 1919 by Robert Gonfreville, a former colonial administrator. Like other colonial industries, it benefited from such agricultural and economic policies as forced cultivation and forced labor. For example, the 1941 Political and Social Report of the Bouaké district noted that the Gonfreville mill employed 800 forced laborers (ANRCI 852, Cercle de Bouaké, *Rapport Politique et Sociale – 1941*, p. 16). Gonfreville became incorporated in 1943, and in the 1950s to early 1970s was controlled by two French groups, Optorg and Texunion. The Company employed 3,285 people in the mid-1970s, and enjoyed government protection for cloth sales in the domestic market, subsidies for seed cotton, tax holidays, and duty-free imports. World Bank, International Finance Corporation, *Report and Recommendation of the President of the Board of Directors on a Proposed Investment in Etablissements R. Gonfreville (ERG), S.A. Ivory Coast* (Report IFC/P-249) (Washington, DC: The World Bank, 1976).
15 Below this length, cotton lost a third of its value. ANRS 1 R 212, Rapport de Mission de Monsieur J. Lhuillier, Directeur de l'IRCT, Le coton en AOF (culture seche) – Les possibilités de production et le programme des recherches envisagées par l'IRCT (1949).
16 *Ibid.*
17 ANRS 1 R 208, Vers. 158, Note sur l'activité de la CFDT en Afrique Occidentale Française, Août 1951.
18 F. Martin, "Compagnie française pour le développement des textiles (CFDT), *Revue Internationale des Produits Coloniaux* 28(1952): 115–20.
19 ANRS 1 R 208, Vers. 158, Circulaire au sujet de la Culture Cotonnière, Abidjan, Fevrier 1951, p. 3.
20 *Ibid.*, p. 4
21 ANRS 1 R 208 d, Vers. 158, Inspect. Gen. de l'Agric., Note sur la production cotonnière en AOF, January 1952.
22 Anon., "Possibilités de développement de la protection cotonnière d'AEF," *Revue Internationale des Produits Coloniaux* 28 (1953), 283: 92–3.
23 ACCCI 24, Procès verbal de la Réunion de la Conférence Annuelle du Coton, 13 October 1951, p. 9.

24 C. Romuald-Robert, *Amélioration de la production cotonnière en Côte d'Ivoire* (Bouaké: IRCT, 1962).
25 ANS-AOF 14 Mi 2033, 2 G 56 (27) Service de l'Agriculture, *Rapport annuel – 1956*, pp. 66–7.
26 Interview with M. Cluchier, Bouaké, 21 March 1992.
27 J. Raingead and C. Romuald-Robert, *Rapport annuel pour la campagne 1955–1956* (Bouaké: IRCT, 1956).
28 SEDES, *Sociologique,* p. 175.
29 Romuald-Robert, *Amélioration.*
30 Interview with Bernadin Ouattara, Korhogo, 23 August 1995.
31 ANRS 1 R 187b, Letter from Lt. Gov. of Côte d'Ivoire to Pres. of Assemblée Territoriale on the subject of Agricultural Service programs, 11 December 1956.
32 Interview with M. Cluchier, Bouaké, 21 March 1992.
33 Aubertin, *Histoire*, 34–7.
34 On 4 September 1947, the former colony of Haute Volta (Upper Volta, now Burkina Faso) was reconstituted, thus reducing the territorial limits of Côte d'Ivoire to its actual borders.
35 L. Gbagbo, *Côte d'Ivoire: économie et société à la veille de l'indépendance (1940–1960)* (Paris: L'Harmattan, 1982), p. 91.
36 H. Labouret, *Paysans d'Afrique occidentale* (Paris: Gaillimard, 1941), p. 29. An "Experimental Committee for Agricultural Mechanization" was created in 1946 by the Department of Agriculture to investigate methods of mechanizing coffee and cocoa cultivation. The committee, however, admitted failure in 1950 and was dissolved in 1952. The major obstacles to mechanization were that the costs were too great for the majority of planters, and the distance between trees was too narrow for machines to pass. The only solution was to uproot trees and establish new plantations – entailing a total upheaval of the colonial economy. Gbagbo, *Economie*, pp. 117–18.
37 The association's revenue came from taxes levied on planters and loggers as well as from contributions made by professional unions and the colonial government. In 1955 the coffee and cocoa marketing board transferred 10% of its total redistributed funds to SIAMO (Syndicat interprofessionnel d'acheminement de la main d'oeuvre)(Aubertin, *Histoire*, p. 32; Gbagbo, *Economie*, pp. 116–22). For details on the labor recruitment activities of SIAMO, see P. Anyang Nyong'o, "Liberal models of capitalist development in Africa: Ivory Coast," *African Development* 3(1978), 194–202.
38 S. Amin, *Le développement du capitalisme en Côte d'Ivoire* (Paris: Editons Minuit, 1967), 30–45, 81–2, 96.
39 SEDES, Région de Korhogo, *Etude de Développement Socio-Economique*: Vol.1, *Rapport Démographique* (Paris: SEDES, 1965), pp. 74–81; Vol. 2, *Rapport Sociologique*, pp. 84–92.
40 The area under African coffee and cocoa increased from 273,000 ha in 1950 to 540,000 ha in 1956. By 1959, the eve of political independence, this area had increased to 733,000 ha. J. Baulin, *La politique intérieure d'Houphouet-Boigny* (Paris: Editions Eurafor-Press, 1982), p. 73.
41 In a 1904 letter to Lieutenant-Governor Marie François Joseph Clozel, Kong district commander Lieutenant Schiffer outlined the main reasons for maintaining the *status quo* ante. Liberated house slaves would find it difficult to pay

head taxes because they lacked access to the means of production; liberation of especially middle-aged house slaves would be psychologically traumatic and disrupt society; European commerce would suffer since household commodity production depended upon the labor of house slaves; and the state risked discontent among slave owners "for no apparent benefit." "For all of these reasons," Schiffer concluded, "it does not seem reasonable to expect that the spontaneous liberation of house captives will produce favorable results for them or for European estates and business." ANRCI MiFm 63–K21, Letter from Lt. Schiffer to Lt. Gov. Clozel, April 1904 on the subject of slavery in the Kong Circle. See Bassett, *Food*, pp. 51–5.

42 Albert Londres also noted the high percentage of *captifs de case* in forced labor gangs. Londres, *Terre d'Ebène*, pp. 47–50; 138.

43 Field notes, Katiali, Côte d'Ivoire, 7 May 1982.

44 This number is derived from the results of a survey conducted by the author in 1982 on migration histories of the members of thirty-eight households in Katiali.

45 Interview with Kaléléna Silué, Katiali, 12 March 1992.

46 Interview with Katienen'golo Silué, Katiali, 13 March 1992.

47 SEDES, Région de Korhogo, *Etude de Développement Socio-Economique*: Vol. 5, *Etudes des budgets familiaux,* (Paris: SEDES, 1965), pp. 24–35.

48 The major factor determining whether one worked in the village chief's large collective field or in one's own lineage's segnon was one's association with the founding matrilineage.

49 Interviews with Bêh Tuo, 3 July 1981; 22 June 1982.

50 SEDES (1965) Région de Korhogo, *Etude de Développement Socio-Economique*: Vol. 2, *Rapport Sociologique* (Paris: SEDES), pp. 36–45.

51 The graph is based on retrospective interviews conducted in the early 1960s by the SEDES research team. Eighty-eight communities were queried about the date of the break-up of their collective fields. The graph shows a peak following the "Houphouet-Boginy Law" ending forced labor, which suggests that "the climate of freedom" of the day enabled some groups to stop working for unpopular village chiefs and lineage heads in collective fields in order to cultivate their own fields. In some communities, change was not as abrupt. For example, the Jula of Katiali (Traorera, Konera, and Diabatera quarters) worked three to five days per week in their quarter's collective field (*forobaforo*) prior to 1946. After that date, they worked just two days per week in these quarter-level fields.

52 SEDES, *Sociologique*, p. 26.

53 Baulin, *Politique*, p. 64.

54 ANRCI, 57, Rapport Politique (1948).

55 *Ibid.*

56 B. Holas, *Les Sénoufo (y compris les Minianka)* (Paris: Presses Universitaires de France, 1957), p. 157.

57 *Ibid.*

58 SEDES, *Sociologique*.

59 Interview with Bêh Tuo, Katiali, 4 April 1982.

60 SEDES, *Sociologique*, pp. 37–39.

61 SEDES, *Sociologique*, pp. 39–40, 57–9.

62 SEDES, *Sociologique*, p. 57.
63 SEDES, *Sociologique*, p. 59.
64 Amin, *Développement*, pp. 99–103.
65 ACCCI, *Comptes-Rendus de CCCI*, 19 February 1956, p. 21.
66 "Allen long-staple" (*G. hirsutum*, race *latifolium*) was originally selected by the American J. B. Allen around 1896. It was commonly grown in the Mississippi delta region until its replacement by more early maturing varieties. Imported into Uganda towards the end of the nineteenth century, *Allen* was introduced into Nigeria in 1912. IRCT, *La culture cotonnière dans le nord de la Côte d'Ivoire* (Bouaké: IRCT, 1976).
67 ANRCI 897a, Essais de 1928 dans les Fermes Cotonnières de Ferkéssédougou et Bouaké, 11 July 1928.
68 RCI, Min. de l'Agric., Service de l'Agric., *Rapport Annuel – 1960 (Partie Technique et Economique)*, p. 57.
69 In 1961/62, 257 ha were cultivated in *Allen* cotton. The Agricultural Service oversaw 73 ha and CFDT monitored 184 ha. RCI, Min. de l'Agric., Service de l'Agric., *Rapport Annuel – 1962 (2eme partie: Partie Technique et Economique)*, p. 57.
70 Yields averaged between 800–900 kg/ha in cotton fields sown in the Bouaké and Béoumi regions.
71 E. Hermann, *Analysis of selected agricultural parastatals in the Ivory Coast*, (Abidjan: REDSO/WA, 1981).
72 Archives de la Chambre de Commerce de Côte d'Ivoire, Dossier 24, Etab. R. Gonfreville, "Programme de Développement de la Culture de Coton," le 25 janvier 1949, p. 2.
73 Groupe de Travail, *Coton*, pp. 204–5; Lele, *et al.*, *Cotton*.
74 P. D. Little and M. J. Watts (eds), "Introduction," p. 9.
75 Y-A Fauré, "Le complexe politico-économique," in Y-A Fauré and J-F Médard (eds.), *Etat et bourgeoisie en Côte d'Ivoire* (Paris: Karthala, 1982), p. 28.
76 The World Bank estimated that between 1987 and 1997, cotton grower incomes were 40% per cent lower due to "stabilization." Amprou, *et al*, "Le coton," pp. 7–8.
77 B. Campbell,"The fiscal crisis of the state: the case of the Ivory Coast," in H. Bernstein and B. Campbell (eds.), *Contradictions of Accumulation in Africa: studies in economy and society* (Beverly Hills, CA: Sage, 1985), p. 277.
78 B. Campbell, "Inside the Miracle: cotton in the Ivory Coast," In J. Barker (ed.), *The Politics of Tropical Agriculture* (Beverly HIlls, CA: Sage, 1984), pp. 143–71. For example, the costs of collecting seed cotton from rural collection points and the expenses incurred in transporting and ginning this cotton formed part of CFDT's declared costs.
79 Bates, *Markets and States*, pp. 12–19.
80 M. Watts and T. Bassett, "Crisis and change in African agriculture: a comparative study of the Ivory Coast and Nigeria," *African Studies Review* 28 (1985), 11–12; E. Lee, "Export-led rural development: the Ivory Coast," *Development and Change* 11 (1980), 607–42; Bates, *Markets and States*, p. 17. Bonnie Campbell's study of the Ivorian textile industry in the 1970s shows a high level of protection and important subsidies in the form of low-cost lint to textile

firms through the intermediary of the Ivorian marketing board. Campbell, "Fiscal," pp. 294–5.

81 Examples of Ivorian state investments in the agricultural sector via these public and parastatal companies included those focused on the development of rice (SODERIZ), sugar (SODESUCRE), and livestock (SODEPRA), as well as cotton (CIDT), coffee and cocoa (SATAMACI), and palm oil (SODEPALM). The rise and fall of the famous SODEs (*société d'état*) is well documented by B. Contamin and Y. Fauré, *La bataille des entreprises publiques en Côte d'Ivoire: histoire d'un ajustement interne* (Paris: Karthala, 1990), and B. Contamin, "Entreprises publiques et désengagement de l'Etat en Côte d'Ivoire: à la recherche des privatisations," in B. Contamin and H. Memel-Fotê (eds.), *Le modèle ivoirien en questions: crises, ajustements, recompositions* (Paris: Karthala and ORSTOM, 1997), pp. 89–107.

82 Fauré, "Le complexe," p. 29.

83 The intermediary position of these firms allows them to profit by inflating their operating costs which are reimbursed by Caistab at the end of the marketing year. In the case of CFDT/CIDT, Bonnie Campbell shows that its declared costs were typically three to four times higher than producer price increases over the period 1972–81. Since these firms are reimbursed by Caistab for their operating costs, they have enjoyed considerable financial benefits to the detriment of rural producers. Campbell, "Inside the Miracle," pp. 158–9

84 This interpretation differs from Bonnie Campbell's view of CFDT's relationship with the state as "preserv[ing] virtually intact the marketing structures inherited from the colonial period." Campbell, "Fiscal crisis," p. 277.

5 MAKING COTTON WORK, 1964–1984

1 Interview with Soro Navigué, Korhogo, 23 August 1995. The coercive role of local administrative authorities in the *Allen* campaign was not confined to the Korhogo region. Subprefects in the Bouaké region also imposed cotton on the local population. SEDES, *Etude Regionale de Bouaké 1962–64*, T. 2, *L'Economie,* (Paris: SEDES, 1965), 166–9.

2 Interview with Soro Navigué, Korhogo, 23 August 1995.

3 *Ibid.*

4 *Ibid.*

5 *Ibid.*

6 Interview with Kaléléna Silué, Katiali, 12 March 1992.

7 Interview with Bernadin Ouattara, Korhogo, 23 August 1995.

8 Interview with Bazoumana Diabaté, Katiali, 28 February 1992.

9 Interview with Katienen'golo Silué, Katiali, 28 February 1992.

10 *Ibid.*

11 J. Peltre-Wurtz and B. Steck, *Les Charrues,* pp. 20–3.

12 *Ibid.*

13 This does not mean that cotton grower incomes were higher as a result of price stability. The World Bank shows that over the period 1987–97, farmer incomes were, on average, 40% lower than world market prices due to the guaranteed price. Amprou, *et al.*, "Etat de Réforms," pp. 7–8.

14 U. Lele and S. Stone, *Population pressure, the environment, and agricultural intensification: variations on the Boserup hypothesis*, MADIA Discussion Paper 4 (Washington, DC: The World Bank, 1989), p. 5.
15 Berry, "Agrarian crisis."
16 For example, a CIDT extension agent in Katiali was responsible in 1984 for monitoring approximately 80 growers and 100 hectares of cotton. A former agent declared that the number of growers and area were not too large and he was able to monitor effectively cotton growing activities *in the field*.
17 Interview with Michel Zahouri, CIDT extension agent, Katiali, 27 Feb 1992.
18 The area in cotton was 35,868 ha in 1970 and 146,400 ha in 1984. Yields increased over the same years from 0.82 t/ha in 1970 to 1.45 t/ha in 1984. See Appendix 1.
19 Hermann, *Parastatals*, p. 144.
20 Inflation indicators are derived from World Bank and IMF estimates of retail price increases in the West African Franc Zone. The inflation rates used to construct Fig. 5.1 were obtained from economists at CFDT's home office in Paris, France.
21 Urea was applied on 80% of the total area.
22 CIDT supplied the plows while farmers obtained oxen from local cattle owners. The immigration of large numbers of Fulbé cattle herders in the 1970s and 1980s increased the number of local cattle for purchase. T. Bassett, "Land use conflicts in pastoral development in northern Côte d'Ivoire," in T. J. Bassett and D. Crummey (eds), *Land in African Agrarian Systems* (Madison: University of Wisconsin Press, 1992), pp. 131–54.
23 The Ministry of Agriculture claims that labor productivity has tripled, while Yves Bigot, Jacqueline Peltre-Wurtz and Benjamin Steck argue that output per active worker has not significantly increased among ox-plow households. Compare RCI, "Bilan diagnostic," with Y. Bigot, "Analyse techno-économique des systèmes de production coton-céréale dans le Nord-Ouest de la Côte d'Ivoire," (Bouaké: Institut de Savanes 1979), and Peltre-Wurtz and Steck, *Les charrues*.
24 Interview with Mr. François Béroud, Director of Rural Development, CFDT, Paris, France, 16 April 1996. Also, see M. Roupsard, *Nord-Cameroun: Ouverture et Développement*, pp. 339–40; 354. A typical ULV sprayer contains a spinning disc atomizer that runs by a battery powered motor. The fine mist is carried by the wind to cotton plants. Since both pesticides and the ULV sprayer were subsidized (100%) in the 1970s and 1980s, the cost of D-size batteries was the most important recurring cost to cotton growers. The sprayers in use in Katiali required eight batteries to spray one hectare. At six applications per hectare, the cost of batteries alone amounted to 2,800 FCFA ($11.20 at pre-devaluation prices).
25 J. M. Munro, *Cotton* (Essex: Longman Scientific and Technical, 1987), pp. 195–6.
26 Interview with Mr. François Béroud, Director of Rural Development, CFDT, Paris, France, 16 April 1996.
27 In 1975, 99.7% of the cotton area was sprayed using the "classic" hand pump

technique. By 1984, peasants were using the ULV technique on 99% of their cotton area. CIDT, *Rapport Annuel d'Activité, Campagne 84/85* (Bouaké: CIDT, 1985), p. 57.

28 A 1981 IMF extended arrangement (3 years) was followed by six standby arrangements between 1984–1991. The World Bank provided three structural adjustment loans (1981, 1983, 1986) and six sectoral adjustment loans between 1989–1991. L. Demery, "Côte d'Ivoire: fettered adjustment" in I. Husain and R. Faruqee (eds.), *Adjustment in Africa: Lessons from country case studies* (Washington, DC: The World Bank, 1994), pp. 72–152.

29 *Ibid.*

30 The rainy season includes those months in which rainfall is greater than 50 millimeters.

31 "Relative droughts" occur when rainfall is less than potential evapotranspiration.

32 P. Fargues, "La dynamique démographique des producteurs du vivrièrs," in *Les cultures vivrières: Element stratégique du développement agricole Ivoirien*, Vol. 1 (Abidjan: CIRES, 1983), 1, pp. 83–104.

33 T. Bassett, "Breaking up the bottlenecks in food-crop and cotton cultivation in northern Côte d'Ivoire," *Africa* 58(1988), 157–161.

34 My observations on cotton growers' efforts to break labor bottlenecks are based on individual and group interviews, field visits, and questionnaires administered to 38 households in 1981–82, 1986, 1988, 1992, and 1995.

35 SEDES, *Agricole*, p. 131.

36 J. M. Kowal and A. H. Kassam, *Agricultural ecology of savanna: a study of West Africa* (Oxford: Clarendon Press, 1978), p. 253.

37 SEDES, *Agricole*, p. 84; *Sociologique*, p. 63.

38 Coulibaly, *Senufo*, pp. 148–50.

39 *Ibid.*, p. 148.

40 Field notes, Katiali, 2 May 1982; Coulibaly, *Sénoufo*, p. 150.

41 Coulibaly, *Sénoufo*, p. 152.

42 Kowal and Kassam, *Agricultural*, pp. 252–7.

43 Bassett, "Political ecology."

44 SEDES recorded 86 different crop associations in the Korhogo region for 1962. For the Millet Zone, 18 associations dominated the cultivated area (80% of the total cultivated area). The top 4 associations are listed in Table 5.4. Associations involving 2 crops accounted for 71% of the total cultivated area. Fields in which 3 crops were intercropped accounted for 8% of the cultivated area. SEDES, *Agricole*, pp. 87–9.

45 Interview with N'golofaga Silué, Katiali, 4 March 1992.

46 *Ibid.*

47 *Ibid.*

48 Y. Bigot, "La culture attelée et ses limites dans l'évolution des systèmes de production en zone de savane en Côte d'Ivoire," *Cahiers du CIRES* 30 (1981),pp. 9–30; Peltre-Wurtz and Steck, *Charrues,* 1979; X. LeRoy, *L'introduction des cultures de rapport dans l'agriculture vivrière sénoufo: le cas de Karakpo* (Abidjan: ORSTOM, 1980); A. Kientz, *Pour une motorisation paysanne; mécanisation et motorisation des exploitations paysannes, Côte d'Ivoire, région nord* (Abidjan: Ministère du Développement Rural, 1985).

49 Eponou's data also show that the average area cultivated per active worker by technique also increased from 0.9 ha in 1962 to 1.27 ha (manual), 1.5 ha (oxen) and 3.75 ha (tractor) respectively. T. Eponou, "*Farm Level Analysis of Rice Production Systems in Northwestern Ivory Coast*," (Ph.D. dissertation, Michigan State University, 1983).
50 SEDES, *Sociologique*, pp. 48–50, 63.
51 Glaze, *Art and Death*, p. 62.
52 *Ibid.*, pp. 153–6.
53 Between 1974 and 1976, Peltre-Wurtz and Steck carried out a time allocation study among 19 agricultural actives in the village of Synofan, 25 km west of Katiali. Peltre-Wurtz and Steck, *Charrues*, pp. 258–9.
54 In 1981 ox-plows were used to plow just 5% of the swamp rice area in Katiali. The area worked by oxen increased to 68% in 1988 and 64% in 1991.
55 Peltre-Wurtz and Steck, *Charrues*, pp. 258–9.
56 P. L. Pingali, Y. Bigot, and H. Binswanger, *Agricultural mechanization and the evolution of farming systems in sub-Saharan Africa* (Baltimore: Johns Hopkins University Press, 1987), pp. 105–8.
57 Bigot, "Culture attelée"; D. Norman, "Farming systems and problems of improving them," in J. M. Kowal and A.H. Kassam, *Agricultural ecology*, pp. 337–40.
58 The average cost amounted to 12,000–15,000 FCFA per hectare.
59 Interview with Donisongui Silué, Katiali, 30 June 1986.
60 The international agribusiness firms, Ciba-Geigy and Rhone-Poulenc, are the main agro-chemical suppliers to village cooperatives.
61 Bigot, "Analyse techno-économique," p. 19.
62 G. Sement, "Etude des effets secondaires de la fertilisation minérale sur le sol dans des systèmes culturaux à base de coton en Côte d'Ivoire: Premièrs résultats en matière de correction," *Coton et fibres tropicales* 35 (1980), 229–48.
63 Peltre-Wurtz and Steck, *Charrues*, pp. 258–9.
64 Berry, *No Condition*, p. 188.
65 Bassett, *Food*, p. 237.
66 Bassett, "Development theory"; M. Demont and P. Jouve, "Evolution d'agroécosystèmes villageois dans la région de Korhogo (Nord Côte d'Ivoire): révision des débats 'Boserup versus Malthus' et compétition versus complémentarité," paper presented at the conference "Dynamiques agraires et construction sociale du territoire," Montpellier, 26–28 April 1999.

6 "TO SOW OR NOT TO SOW": THE EXTENSIFICATION OF COTTON, GENDER POLITICS, AND RURAL MOBILIZATION, 1985–1995

1 USDA, *Agricultural Statistics* (Washington, DC: Government Printing Office, 1988).
2 At the risk of simplifying a complex marketing structure, it should be noted that cotton sales are denominated in US dollars on the world market. For the early part of 1985, the dollar was strong in world markets, exchanging at the rate of 10.1 French Francs per dollar. However, at about the same time that cotton prices plummeted, the strength of the dollar also dropped steeply,

which only served to exacerbate the Ivorian cotton crisis. Even more complicated is CFDT's cost accounting system. Bonnie Campbell shows that the French firm systematically inflates its industrial and commercial costs to achieve high profit margins. Campbell, "Inside the miracle," pp. 143–71.

3 République de Côte d'Ivoire, Min. de l'Agriculture et des Eaux et Forets, Min. de l'Economie et des Finances, *Prêt d'ajustement sectoriel de l'agriculture: Etude des conditions d'équilibre de la filière coton; Preparation d'un contrat-plan entre l'état et la CIDT Provisoire*, Novembre 1990 (Abidjan: Direction et Controle des Grands Travaux), p. ii.

4 "France/Africa: a devalued relationship," *Africa Confidential* 35(2), January 1994, p. 7.

5 As a result of the devaluation, the US dollar/FCFA exchange rate doubled in favor of the dollar. Flush with dollars from world market sales, the state was in the position to increase the proportion of value added to producers. Instead, the part captured by peasant producers fell from 25% to 15% after the devaluation. D. Herbel, *La compétitivité du coton ivoirien*, Document de travail (Montpellier: CIRAD, 1995), p. 76.

6 Two glandless varieties (ISA BC2 and BC4) were grown in 1984 on 23,700 hectares (16% of the northern cotton area). However, lower than expected gin yields led CIDT to abandon this experiment the following year (Groupe de Travail, *Coton*, p. 150). The seeds of the classic *Allen* and ISA varieties are toxic and can only be consumed by cattle (CIDT, *Rapport Annuel, Campagne 1990/91* [Bouaké: CIDT, 1990]).

7 Lele, *et al.*, "Cotton," p. 10.

8 It is also worth noting that extension agents failed to instruct growers sufficiently about the exigent qualities of GL7, particularly about the prospects of low yields if CIDT's rigorous cropping calendar was not followed closely.

9 The General Manager of CIDT explained why GL7 was ultimately eliminated from the HYV program in 1996. "The introduction in 1991–92 of the glandless variety, which was not very hardy but very interesting in terms of its productivity and gin yields, and its diffusion throughout the northern CIDT region, was not well received by producers who found it too unstable, difficult to grow, and sensitive to pests. Its rejection led to a decline in cultivated area and yields, and we repeatedly found ourselves in the completely abnormal situation where recorded yields were higher in CIDT's central and western regions in comparison to the north. Faced with this problem and not wishing to compromise the resumption of production, CIDT decided to put an end to the glandless experience at the end of the 1995–96 campaign." S. Coulibaly, "Le rôle de la CIDT est primordial: entretien avec M. Samba Coulibaly, directeur général de la CIDT," *Coton et Développement* 21 (1997), p. 6.

10 H. Bernstein, "Notes on peasants and capital," *Review of African Political Economy* (1979), p. 427.

11 For example, the undeclared cotton area ranged from 25% to 40% in northern Guinea during the early 1990s (F. Geay and C. Konomou, *Pratiques paysannes en culture cotonnière*, Projet de développement rural de la Haute-Guinée, Suivi-Evaluation [Conakry: PDRHG, 1993], pp. 3–4). In Mali, the percentage of undeclared area varied among regions. The national average amounted to 10%. (F. Giraudy, *Annuaire statistique 94/95: résultats de l'enquête agricole permanente* [Sikasso: CMDT, 1995], p. 7.) CFDT rural development experts believed

that this "undeclared" area was evidence of extensification (interview with François Geay, CFDT, 16 April 1996, Paris, France).

12 M. Fok, *Le développement du coton au Mali par analyse des contradictions: les acteurs et les crises de 1895 à 1993*, Unité de Recherche Economie des Filières No. 8 (Montpellier: CIRAD, 1994), p. 178.
13 The recommended dosage is 200 kg of NPK and 50 kg of urea per hectare.
14 Interview with CIDT monitor, Katiali, 9 August 1995.
15 Fertilizer underdosage is widespread in Northern Cameroun (J-C Sigrist, "Pratiques paysannes et utilisation des intrants en culture cotonnière au nord-Cameroun," Memoire de Fin d'Etudes, Institut Supérieur Technique d'Outre-Mer [1992], p. 107), Burkina Faso (P. Lendres, "Pratiques paysannes et utilisation des intrants en culture cotonnière au Burkina Faso," Mémoire de Fin d'Etudes, Centre National d'Etudes Agronomiques des Régions Chaudes [1993], p. 53), Senegal (J-R Cuzon, "Réalités des pratiques paysannes en matière d'utilisation des intrants sur le coton au Sénégal," Mémoire Diplôme d'Agronomie Tropicale, Ecole Nationale Supérieure des Sciences Agronomiques Appliquées, [1993], p. 51), Benin (C. Colnard, "Pratiques paysannes et utilisation des intrants en culture cotonnière au Benin," Mémoire de Fin d'Etudes, Ecole Nationale Supérieure d'Horticulture de Versailles [1995], p. 74), and Guinée (Geay and Konomou, *Pratiques*, pp. 9–12).
16 Coulibaly, "Rôle."
17 CIDT, *Rapport Annuel, Campagne 1994/95* (Bouaké: CIDT, 1995), p. 2.
18 For a more detailed discussion of women cotton growers in Katiali, see T. Bassett,"Migration et féminisation de l'agriculture dans le nord de la Côte d'Ivoire," in C. Meillassoux and B. Schlemmer (eds.), *Déséquilibres démographiques, déséquilibres alimentaires* (Paris: Etudes et Documentations Internationales, 1991), pp. 219–45.
19 Interview with Gniré Tuo, Katiali, 10 August 1995.
20 Interview with Tiékundôh Silué, Katiali, 29 June 1998.
21 Some Senufo Muslims considered Friday a day of prayer and rest. Also, many places in rural areas have designated rest days when farming should not take place to appease the bush spirits of that locale. Friday and Monday are commonplace rest days (*segui tiandin*).
22 J. Carney and M. Watts, "Manufacturing dissent: work, gender, and the politics of meaning in peasant society," *Africa* 60 (1990), 207–41; D. Moore, "Marxism, culture, and political ecology," in R. Peet and M. Watts (eds.), *Liberation Ecologies* (London: Routledge, 1996), pp. 125–47.
23 Pursell defines the *ristourne* as a supplementary payment made to growers by cotton companies based either on company profits during the preceding year or on a formula linked to world prices of cotton lint. G. Pursell, "Cotton Policies in Francophone Africa," manuscript, World Bank, May 21, 1998, p. 3.
24 GVC de Katiali, *Campagne 1991/92*.
25 Interview with Mr. Zéha Silué, 14 February 1992.
26 Bernstein, "Peasantries," pp. 428–29.
27 Interview with Adama Koné, Katiali, 7 March 1992. Mr. Koné stated that cotton was marketed by both CIDT and the GVC for a few years in Katiali. As he learned about the process of cotton marketing during this transition period, he came to doubt that CIDT systematically cheated peasants.
28 This *local* demand for an alternative marketing organization illustrates the

endogenous dimension to this village-level institutional innovation. This observation contrasts with that of Xavier Le Roy, who describes the GVC structure as "completely exogenous." As argued in chapter 1, such innovations should be seen as emerging from the interactive effects of exogenous and endogenous forces under changing social and economic conditions. X. Le Roy, "Où la culture cotonnière rénove" p. 203.

29 In 1994, the commission amounted to 4002 FCFA ($15) per metric ton.

30 In contrast to the large number of cotton growers participating in the GVC structure in the savanna zone, just a third of coffee and cocoa producers adhere to the GVC structure in the forest region. Conflicts among members and non-members over how to spend the *ristourne* is one source of tension making the GVC less effective in both economic and political life in the forest area. See S.Y. Affou, "Renforcement des organisations paysannes et progrès agricole: obstacles ou atouts pour le progrès agricole," in B. Contamin and H. Memel-Foté (eds.), *Le Modèle ivoirien en questions: crises, ajustements, recompositions* (Paris: Karthala and ORSTOM, 1997), pp. 555–71; and J. Widner, "The discovery of 'politics': smallholder reactions to the cocoa crisis of 1988–90 in Côte d'Ivoire," in T. M. Callaghy and J. Ravenhill (eds.), *Hemmed in: responses to Africa's economic decline* (New York: Columbia University Press, 1993), pp. 279–331.

31 CIDT, *Rapport Annuel, Campagne 89/90* (Bouaké: CIDT, 1990), p. 100.

32 Bankruptcy forced the closure of BNDA in Octobre 1991. In the central cotton growing areas where animal traction is less developed, CIDT extended credit to cotton growers to purchase oxen. CIDT, *Rapport Annuel, Campagne 91/92* (Bouaké: CIDT), p. 48.

33 Interview with Mr. Mamadu Dosso, URECOS-CI, Korhogo, 2 February 1992.

34 *Ibid.* URECOS-CI opened its office in November 1991, the same month as the first cotton strike. URECOS represents GVC members at local, regional, national, and international levels, provides services to GVCs ranging from negotiating with input suppliers to training GVC leaders, and seeks to play a more active economic role in cotton marketing. URECOS's 1993–94 annual report lists a membership of 759 village cooperatives of which 300 were organized into 34 subregional unions. The report indicates that many of these GVCs had not paid their dues and that considerable tension existed between URECOS and some of its members over URECOS financial management and bargaining record with CIDT. URECOS-CI (1995) *Rapport d'activités, Exercice 1993/94* (Korhogo: URECOS-CI); URECOS-CI (1995) *Rapport moral, Exercice 1993/94* (Korhogo: URECOS-CI).

35 Affou, "Renforcement," p. 568.

36 Interview with Brahima Coulibaly, M'Bengué, 29 February 1992.

37 Demery, "Fettered," p. 98.

38 USDA, Foreign Agricultural Service, "CIDT Privatization," AGR IV5010, 1995, p. 1.

39 The CIDT committee that sets producer prices is named "Comité paritaire de fixation des prix du coton." Cotton growers are also represented on CIDT's Board of Directors, and on a company committee that sets prices for cotton inputs like fertilizers and pesticides named "Comité de pilotage de sous filière intrant CIDT".

40 Interview with Brahima Coulibaly, M'Bengué, 29 February 1992.

41 The heads of GVC unions had already been meeting since the first price cut in 1990 to create an alternative organization (URECOS-CI) to compete with CIDT in input provision and marketing. In addition to the 1990 price cut, the GVC membership was unhappy about higher fertilizer prices, the increasing proportion of low-grade classified cotton, peasant indebtedness, and CIDT's success in muzzling DMC. CIDT, *Campagne 89/90*, pp. 101–2; RCI, *Prêt d'ajustement*, p.14.
42 This fertilizer price reduction amounted to a savings of 4,775FCFA/ha ($19). URECOS-CI, *Rapport moral: Exercice 1993/94* (Korhogo: URECOS-CI, 1995), p. 5.
43 Interview with Patrick Bisson, CIDT-Bouaké, 22 January 1992.
44 CIDT, *Rapport d'Activité de la Campagne Agricole, 1993–1994*, (Bouaké: CIDT, 1994), pp. 14, 17.
45 H. Solo (1995), "Objectif immédiat: 300.000 tonnes," *Fraternité Matin*, 9 July 1995, p. 4.
46 Herbel, *La compétitivité.*
47 World Bank, *Côte d'Ivoire: Revue du secteur agricole*, Document du travail préparé pour l'atelier de la revue du secteur agricole (7–10 novembre 1994), p. 23.
48 Herbel, *La compétitivité,* p. 119.
49 Solo, "Objectif," p. 4. According to an agricultural attaché report issuing from the American Embassy in Abidjan, this subsidy amounted to 200 FCFA per kilogram of cotton fiber (USDA, Foreign Agricultural Service, "Cotton Annual," AGR IV6012, 1995, p. 2). The subsidized cotton sales continued in 1996–97 when CIDT sold 19,000 metric tons of fiber to local textile mills at a price of 665 FCFA per kilogram of cotton fiber (USDA, Foreign Agricultural Service, "Annual Report," AGR IV7010, 1995, p. 6).
50 *Fraternité Matin*, "Objectif," p. 4.
51 Demery, "Fettered," pp. 128–29; N. Van de Walle (1991), "The decline of the Franc Zone: monetary politics in Francophone Africa," *African Affairs* 90 (30), 383–406.
52 Herbel, *La compétitivité,* p. 153.
53 CIDT, *Campagne 94/95*, p. 41.
54 J-L. Chaleard, *Temps de vivres, temps des villes* (Paris: Karthala, 1996).
55 Female informants explained that the principal reasons behind their decision to stop growing cotton were as follows: conjugal conflicts over access to family labor; an inability to mobilize labor during critical moments in the agricultural calendar (e.g. plowing fields, sowing and pesticide spraying periods); their dependence on official growers to market their cotton; the high price of inputs; and the low yields of GL7.
56 The cooperative "Womiengnon de Niofoin" bought cashews for 250 FCFA/kg.
57 CIDT, *Rapport Annuel, Campagne 92/93* (Bouaké: CIDT, 1993), p. 3.
58 Interview with Mr. François Béroud, Directeur du Développement Rural, la Compagnie Française pour le Développement des Fibres Textiles (CFDT), Paris, 16 April 1996.
59 Bassett, "Political ecology"; Bassett, "Hired herders."
60 According to Affou, the Front Populaire Ivoirien (FPI), a main opposition party, "created" COOPAG-CI (Affou, "Renforcement," p. 568). Widner

reports that the FPI was also active in organizing the first party congress of the Syndicat Nationale des agriculteurs de Côte d'Ivoire (SYNAGCI), where FPI leader Laurent Gbagbo ultimately informed his audience that unions should be independent and not associated with political parties. Gbagbo was quoted as saying: "There is no such thing as PDCI cotton or FPI cotton. There is just cotton." (Widner, "Discovery," p. 323). The creation of crop damage committees along party lines was another manifestation of multi-party politics influencing community institutions. For example, in 1992 there were two crop damage committees in Katiali, one associated with the ruling PDCI party, the other with the FPI. There was a noticeable increase in farmer compensation as a result of the rivalry between these competing groups (Basset, "Hired herders").

61 URECOS-CI attracted the largest number of growers as indicated by the percentage of cotton marketed through their structure. For the 1995/96 season, URECOS-CI members produced and marketed 79% of the total harvest, while COOPAG-CI and CEACI's shares were 16% and 5% respectively. CIDT, *Rapport Annuel, Campagne 95/96* (Bouaké: CIDT, 1996), p. 43.

62 As the responsibilities of GVCs increased, so did the number of marketing commissions. In addition to the one received for marketing cotton, cooperatives now receive commissions for distributing fertilizers and pesticides.

63 Some ousted members from Katiali's GVC sought to join cooperatives in neighboring communities (e.g. M'Bengué) to obtain agricultural credit.

64 The number of cooperatives marketing cotton in the Korhogo region alone increased 43% from 171 in 1994/95 to 245 in 1995/96. CIDT, Direction Régionale, Korhogo, *Bilan Opération: Commercialisation and Résultats, Campagne: 1995–1996*, (Korhogo: CIDT, 1996), p. 1.

65 In contrast to the forest region, where significant social differentiation exists among farmers, cotton growers are characteristically small farmers. In 1995, the average area cultivated was 1.5 ha.

66 The sums of money are not insignificant in light of low farmer incomes. The supplementary payment (*ristourne*) to village level cooperatives for the marketing and loading of seed cotton into transport containers amounted to $2,489,000 at an exchange rate of 550 FCFA to one US dollar. They earned an additional $105,552 in commissions for managing the distribution of fertilizers for CIDT. CIDT, *95/96 Campagne*, p. 43.

67 "Mouvement coopératif: les OPA de la filière cotonnière," *Coton Magazine*, 3 (May) 1998, pp. 14–15.

68 Jennifer Widner offers an excellent analysis of the domestic political and economic forces that led to the emergence of a "competitive" multi-party political system in Côte d'Ivoire in 1990. Widner, "Political reform."

69 A spokesperson for COOPAG-CI believed that his organization was a target of destabilization by unnamed agents because "they absolutely do not want to see peasants become professional (farmers)." Interview with Mr. Martin Yao, *Coton Magazine*, (May) 1998, p. 18.

70 The felicitous phrase comes from Michael Watts, "Development II." Despite the restructuring and privatization of public enterprises and parastatals in the 1980 and 1990s, Contamin concludes that the nature and direction of privatization has been "neither massive nor linear" (Contamin, "Désengagement," p.

98). Widner draws a similar conclusion although interprets this restructuring as leading to a squeeze on elite incomes and ultimately political reform. See Widner, "Political reform," pp. 61–64.

71 In 1994 CFDT was a minority share holder in the cotton companies of the following franc zone countries: Burkina Faso – Sofitex (34%); Cameroon – Sodecoton (30%); Central African Republic – Sococa (34%); Côte d'Ivoire – CIDT (30%); Mali – CMDT (40%); Senegal – Sodéfitex (20%), and Chad – Cotonchad (17%). It also held equity shares in the cotton companies of other African countries such Madagascar-Hasyma (29.4%), Zaire-Codénord (37.5%), as well as in Spain-MASA (50%).

72 M. Fichet, "Côte d'Ivoire: Derrière le concept de privatisation, le démantlement d'une filière," *Coton et Développement* 22 (1997), pp. 2–3; S. Dupont "La guerre du coton noir: Américains et Français s'opposent en Afrique," *Les Echos*, 16–17 February 1996, pp. 50–51.

73 CFDT's financial interests are linked not only to its minority share holdings in West African cotton companies, but also to its technical consultancy contracts and cotton trading activities of its subsidiary, La Compagnie Cotonnière (COPACO). COPACO, based at CFDT headquarters in Paris, markets nearly half of the cotton produced in French West Africa. Pursell, "Cotton policies," p. 1.

74 The six gins went on the market in two blocks in the spring and summer of 1998: the north-east block contained the Korhogo I, Korhogo II, and Ouangolodougou factories with a total ginning capacity of 119,000 tons; the north-west block consisted of the Boundiali I, Boundiali II, and Dianra factories with a ginning capacity of 101,000 tons. The north-west block went to Ivoire Coton, owned by Industrial Promotion Services of the Aga Khan group, while the north-east block was sold to La Compagnie Cotonnière Ivoirienne, owned by the Swiss company Aiglon. The four gins located in the center of the country in Séguéla, Bouaké, Zatta, and Mankono (capacity 103,000 tons), were not put up for sale. They were held by the "New CIDT" with capital shares held by CFDT (30%) and the Ivorian government (70%).

75 USDA, Foreign Agricultural Service, *Côte d'Ivoire: Cotton Annual 1999*, GAIN Report IV9011, June 5, 1999, pp. 1–2.

76 Fichet, "Derrière," p. 2.

77 *Ibid.*

78 Solo, "Objectif," p. 4.

7 CONCLUSION

1 On the maize revolution in sub-Saharan Africa see D. Weiner, S. Moyo, B. Munslow, and P. O'Keefe, "Land use and agricultural productivity in Zimbabwe," *J. of Modern African Studies* 23 (1985), 251–85; D. Byerlee and C. Eicher, *Africa's emerging maize revolution* (Boulder: Lynne Rienner, 1997).

2 R. Launay, "Stereotypic vision: the 'moral character' of the Senufo in colonial and postcolonial discourse," *Cahier d'Etudes Africaines*, 49 (1999), 154: 271–92.

3 Bates, *Markets*, pp. 11–29.

4 Michelis, "Pourquoi," p. 32.

5 *The American Heritage Dictionary of the English Language*, 3rd edition, (Boston: Houghton Mifflin Company, 1992), p. 1545.

6 For example, the 1995/96 indicators show that Odienné's cotton yields averaged 885 kg/ha and just 30% of the cotton area was cultivated by oxen. Boundiali recorded an average yield of 1107 kg/ha and 73% of the area worked by ox-plows. CIDT, *Rapport Annuel, Annexes Statistiques, Campagne 1995/1996* (Bouaké: CIDT, 1996), pp. 4–5.

7 Le Roy, "Où la culture."

8 M. Demont and P. Jouve, "Evolution d'agroécosystèmes villageois dans la région de Korhogo (Nord Côte d'Ivoire): révision des débats 'Boserup versus Malthus' et compétition versus complémentarité," paper presented at the symposium "Dynamiques agraires et construction sociale du territoire," Montpellier, 26–28 April 1999.

9 Campbell, "Inside the Miracle," pp. 158–59; 166.

10 The previous record (291,000 tons) was set in 1988/89.

11 U.S. Department of Agriculture, Foreign Agricultural Service, *Cotton Annual – 1999*, p. 2.

12 Farmers received an additional 5 FCFA/kg for their record-breaking performance in 1998/99. Despite the price hike, the real purchase price stagnated around 99 FCFA/kg between 1988 and 1997. As a World Bank report concluded, "the benefits of the [1994] devaluation on FCFA prices were not transmitted to producers." J. Amprou, L. Gbeli, and J. McIntire, "Le coton en Côte d'Ivoire: état des réformes," mimeograph, Banque Mondiale (3 juin 1998), p. 1.

13 High humidity levels at harvest time, poor sorting, and the yellowish tint of the N'TA variety resulted in just 58% of the crop being classified as first grade. *Ibid.*, p. 3.

14 Anon.,"La zone franc prouve son dynamisme en atteignant un nouveau record de production," *Afrique Agriculture*, 260, June 1998, p. 18.

15 The three gins were located in M'Bengué, Korhogo, and Odienné. The M'Bengué and Odienné gins were financed by the two companies that were successful in purchasing CIDT's northern gins during the 1998 privatization sale (the Compagnie Cotonnière Ivoirienne, owned by the the Swiss firm Aiglon, and Ivoire Cotton, owned by the Agha Khan group). The Korhogo gin was financed by the farmer organization, URECOS-CI, and the French Louis Dreyfus Group. The farmers' organization had sought approval from the Ministry of Agriculture to build two gins, one in M'Bengué and one in Korhogo. They argued that existing processing capacity was insufficient and partly responsible for delays in collecting seed cotton held in rural areas. Cotton quality deteriorated, resulting in a large proportion (42% in 1998/99) of it being classified as second-grade by gin operators. USDA, "Côte d'Ivoire, Cotton Producer Price," Gain Report IV8022, December 11, 1998, p. 1; USDA,"Côte d'Ivoire, *Cotton Annual-1999*, Gain Report IV9011, p. 3.

16 *Afrique Agriculture*, 271, June 1999.

17 H. Solo, "Filière coton: savoir raison garder," *Fraternité Matin*, 24–25 décembre 1999, p. 10; A. Koné, "Coton: la commercialisation reprend," *Le Jour* 1486, 21 janvier 2000.

18 URECOS-CI's membership included 1,200 cooperative groups and 50 cooperative unions.

19 Personal communication, Antoine Anzele, US Department of Agriculture, Foreign Agricultural Service, Abidjan, Côte d'Ivoire, 13 January 2000. The new gin, the third to be located in Korhogo, was to have a capacity of 30,000 metric tons and be operational for the 2000/01 marketing year.
20 One should really speak of a partial privatization, since CIDT continued to hold a 30% share in the six gins bought by the foreign owned companies. It was to hold its shares for a three-year period before selling them off to private investors. CIDT also continued to control the remaining four gins located in the center of the country.
21 Interview with Bema Koné, Katiali, 9 July 1995. Bema cultivated 5 ha of the demanding GL 7 cotton variety in 1994/95. His yield, 564 kg/ha, was a quarter of CIDT's optimistic projections. With the high-grade cotton price set at 160 FCFA/kg, he earned 451,520 FCFA. However, he had to reimburse the GVC cooperative 415,000 FCFA for the cost of fertilizer, pesticides, and herbicides he obtained on credit. His net income that year was 36,520 FCFA, less than $70.
22 For an excellent summary and critique of this literature, see Berry, "Food crisis," pp. 59–112.
23 On the impact of cotton growing on soils see F. Mahop and E. Van Ranst, "Côut de l'épuisement du sol en zone cotonnière camerounaise: impact sur l'environnement," *Tropicultura* 15 (1997), pp. 203–8; G. Sement, "Etudes des effets secondaires de la fertilisation minérale sur le sol dans des systèmes culturaux à base de coton en Côte d'Ivoire: premiers résultats en matière de correction," *Coton et Fibres Tropicales* 35 (1980), pp. 229–248; L. Gray, "Land degradation," pp. 128–181. Pest resistance is of increasing concern in the cotton-growing areas of West Africa. The resurgence of *Helicoverpa armigera* since 1996 in Mali, Burkina Faso, Benin, and northern Côte d'Ivoire is strongly believed to be linked to its resistance to pyrethrum-based insecticides. "Mieux protéger le contonnier contre *Helicoverpa armigera*," *Afrique Agriculture* 271 (1999), pp. 64–65. See also, M. Vaissayre, "La résistance aux pesticides des ravageurs: une menace pour la culture contonnière en Afrique?" *Afrique Agriculture* 260 (1998), pp. 65–66. François Béroud, CFDT's Director of Rural Development, provides an overview of some of the main cotton-environmental issues. His conclusion that environmental concerns are exaggerated is based more on his political role in defending CFDT against its detractors during the privatization debates of the 1990s than on sound research. F. Béroud (1999), "Coton et Environnement: des idées reçues dont il faut se méfier," *Afrique Agriculture* 271 (1999), pp. 59–61.
24 P. Richards, *Coping with hunger: hazard and experiment in an African rice-farming system* (London: Allen and Unwin, 1986), p. 148.
25 Price stabilization did not necessarily mean that peasant "security" was strengthened. In fact, World Bank economists calculated that peasant incomes would have been higher if prices fluctuated with world market prices. They argue that over the period 1987/88–1997/98, growers lost 38.1 billion FCFA, or 40% of the average value of the guaranteed producer price, because of their "protection" from world market price flucuations. Amprou, *et al.*, "Le coton," 7–8.
26 Interview with Bema Koné, Mariam Ouattara, and Fatima Koné, Katiali, 5

July 1998. Because of her status as senior wife, Bema's third spouse, Sita Koné, was allowed to cultivate a cotton field (0.5 ha) in 1998.

27 Bates, *Essays*, p. 131.

28 Bates, *Markets*, pp. 121–32. Jennifer Widner expands Bates's basic argument by suggesting that "favorable" agricultural policies in Côte d'Ivoire are linked to the stakes held by the political elite in both agriculture and agro-industry. She argues that agribusiness managers' concerns for a steady supply of high-grade products have led them to lobby for nominal price increases to maintain peasant interest in agriculture. Although this may be the case for the palm oil industry, it has not been the case for the cotton textile industry. Its managers have successfully lobbied for subsidized cotton lint and thus effectively taxed peasant cotton growers. Moreover, national cotton production far exceeded local fiber needs during the 1980s and 1990s, which suggests that the textile industry was less concerned with insufficient cotton supplies and quality than with securing cheap cotton. J. Widner, "The origins of agricultural policy in Ivory Coast, 1960–86," *J. of Development Studies* 29 (1993): 49–52.

29 M. Bratton and J. Bingen, "Farmer organization and agricultural policy in Africa – Introduction," *African Rural and Urban Studies* 1:1 (1994), p. 9.

30 This is not to imply that the move towards democratization was directed or otherwise inspired by aid donors. Jennifer Widner persuasively shows that domestic political and economic forces (e.g. the repercussions of the fiscal crisis on patrimonial politics), combined with the ruling party's control over the media, were very important in the political reforms leading to multi-party competition in Côte d'Ivoire. Widner, "Political reform."

31 For a comparative case study of the challenges facing farmer organizations in Mali, see J. Bingen, "Agricultural development policy and grassroots democracy in Mali: the emergence of Mali's farmer movement," *African Rural and Urban Studies* 1 (1994), pp. 57–72.

Bibliography

ARCHIVAL SOURCES

ARCHIVES NATIONALES DE LA RÉPUBLIQUE DE CÔTE D'IVOIRE (ANRCI)

ANRCI 1 EE 79(5)b, Cercle de Kong, Poste de Korhogo, Rapport mensuel, November 1915.

ANRCI 1 QQ 81, Situation agricole de Korhogo, 1909.

ANRCI 1 QQ 81b, Cercle de Korhogo, Rapport économique, 23 April 1912.

ANRCI 1 QQ 81d, Cercle de Kong, Poste de Korhogo, Note succinte sur la situation agricole, 6 December 1916.

ANRCI 1 RR 14, Correspondance au sujet de l'intensification de la production agricole pour ravitailler la metropole – 1916–1918; letter from Governor General, 7 June 1917.

ANRCI 1 RR 63, Rapport de Mon. Waddington sur l'état actuel et de l'avenir de la culture cotonnière en Afrique Occidentale, 3 February 1924.

ANRCI 1 RR 63a, Situation de la culture de coton en Côte d'Ivoire, Rapport sur la culture de cotonnière: plan de campagne, 1913.

ANRCI 1 RR 63c, Correspondances relatives au plan de campagne de culture cotonnière dans les Cercles N'zi-Comoé, Baoulé, Ouorodougou, Tagouanas et Kong: letter from Lieutenant Governor of Côte d'Ivoire to Administrators of Circles of 31 Janaury 1918; letter from the Chief of the Agriculture Service to Lieutenant Governor of Côte d' Ivoire of 16 March 1917.

ANRCI 1 RR 63m, Rapport de tournée par L. Leraide, Chef du Service de l'Agriculture, 15 March 1915.

ANRCI 1 RR 63d, Correspondances relatives à la campagne cotonnière et aux maladies du cotonnière, 1917.

ANRCI, 1 RR 64b, Campagne cotonnière en Côte d'Ivoire, 1925–1927.

ANRCI 1 RR 65, Correspondance relative aux champs d'essais de coton dans les cercles – 1927–1928; télégramme officiel de l 'Inspecteur, Service des Textiles, 30 April 1928.

ANRCI, 1 RR 67a, Correspondances relatives à l'envoi des semences, à l'achat des semences et à l'intensification de la culture du coton dans le cercle de N'zi-Comoé, 14 February 1925.

ANRCI 1 RR 67b, Circulaires relatives au développement de la culture du coton dans le cercle de N'zi-Comoé, 7 May 1925.

ANRCI 1 RR 67c, Correspondance relative à la vente, à la production du coton et aux materiels textiles, 1912–31.

ANRCI 1 RR 67d, Rapport des tournées effectuées par l'Ingenieur des Etages dans la zone de culture d'Ishan en vue de l'étude des questions cotonnière et principalement des maladies, 1936.

ANRCI 57, Rapport politique, 1948.

ANRCI 426, Rapport des tournées effectuées du 25 avril au 30 mai 1942 dans le Cercle de Korhogo par l'Administrateur du Cercle, 3 June 1942.

ANRCI 598, Letter from the High Commissioner to the Minister of Overseas France, a/s Qualité et prix coton Côte d'Ivoire, 6 January 1949.

ANRCI 598, Letter from Governor Pechoux to the High Commissioner of French West Africa, 5 February 1949.

ANRCI 598, Letter from the Governor of Côte d'Ivoire to the Governor General of AOF, a/s plan de production cotonnière, 24 Sept 1938.

ANRCI 812, Association Cotonnière Coloniale, Status.

ANRCI 813, Cercle de Kong, Campagne Cotonnière 1935–36.

ANRCI 840 Service des Textiles, Correspondance avec les cercles du nord; rapport de l'Ingenieur des Etages sur son tournée effectuée dans la région sepentrionale de la Côte d' Ivoire, 28 October, 1935.

ANRCI 840a, Service des Textiles, Correspondance avec les cercles du nord; letter from Lieutenant Governor of Côte d'Ivoire to Administrator of Korhogo Circle on the subject of peanut cultivation, 15 April 1929.

ANRCI 840b, Rapport de l'Ingenieur des Etages sur son tournée effectuée dans la région septentrionale de la Côte d'Ivoire, 28 October 1935.

ANRCI 861, Cercle de Kong, Situation agricole, 1935.

ANRCI 897a, Essais de 1928 dans les Fermes Cotonnières de Ferkéssédougou et Bouaké, 11 July 1928.

ANRCI 897b, Service des Textiles de Bouaké, Rapport du ferme cotonnière de Bouaké, 1932.

ANRCI 1115, Service des Textiles, 1928–37.

ANRCI 2704, Rapport politique, Année 1958, par D. Pinelli.

ANRCI 2896, L'affaire Monsignor Diss, relative à la pretendu famine dans le Cercle de Korhogo, 12 September 1931.

ANRCI 2990, Inspection des Affaires Administratives, Rapport à Monsieur le Governor de la Côte d'Ivoire a/s des usages commerciaux particuliers au Cercle de Korhogo et difficultés qui sont produits entre les maisons de commerces et l'Administrateur M. Lalande, 26 January 1931.

ANRCI 3039, Letter from the Head of the Boundiali Subdivision to the Commander of the Korhogo Circle, 6 August 1943.

ANRCI 3039, Letter No. 2084, from the Commander of the Korhogo Circle to the Governor of Côte d'Ivoire, 14 August 1943.

ANRCI 3442, Campagne cotonnière 1928.

ANRCI MiFm 5G 67, Letter of Lieutenant Governor Angoulvant on the political, economic and administrative situation of Côte d'Ivoire, 26 November 1908.

ANRCI, Monographes, Essai de Monographie du Cercle de Korhogo, par Maurice Delafosse, 1905.

ARCHIVES DE LA CHAMBRE DE COMMERCE DE CÔTE D'IVOIRE (ACCCI)

ACCCI 24, L. Chabrand, Note sur le coton en Côte d'Ivoire, Sept. 1944.
ACCCI 24, Etablissements R. Gonfreville, Bouaké, Côte d'Ivoire, Culture de Coton en Côte d'Ivoire, Oct. 1951.
ACCCI 24, Procès verbal de la Réunion de la Conférence Annuelle du Coton," 13 Oct. 1951.
ACCCI 324, CFDT, Conditions du développement de la production cotonnière en Côte d'Ivoire," May 1952.
ACCCI, *Comptes-Rendus de CCCI*, 19 Feb. 1956.

ARCHIVES NATIONALES DE LA RÉPUBLIQUE DU SÉNÉGAL, GOVERNMENT-GENERAL AOF SERIES (ANRS)

ANRS 1 G-146, Mission du Captain Quiquandon dans le Kénédougou, 1890–91.
ANRS 1 R 48a, La culture de coton dans les différentes colonies de l'AOF, 1903–30.
ANRS 1 R 48, Vers. 158, Enquête sur la culture du coton, 27 August 1918.
ANRS 1 R 48, Vers. 158, Letter from Governor General to the Governor of Côte d'Ivoire, 5 Sept 1918.
ANRS 1 R 48, Vers. 158, Letter from Governor General Angoulvant to Governor Antonetti, 31 October 1918.
ANRS 1 R 108a, Circulaire sur l'intensification de la production agricole et pastorale en Côte d'Ivoire, 4 April 1931.
ANRS 1 R 108c, Cercle de Kong, Rapport sur l'exécution de la première tranche du programme agricole quinquennal, 5 July 1932.
ANRS 1 R 108d, Letter from the Governor of Côte d'Ivoire to Governor General of AOF, 11 August 1932.
ANRS 1 R 108e, Letter from Governor General de Coppet to the governors of the French Soudan, Sénégal, Côte d'Ivoire, and Dahomey on the subject of bush cotton, 19 February 1938.
ANRS 1 R 108f, Programme de développement de la culture cotonnière en AOF, 20 August 1938.
ANRS 1 R 125, Recherche et selection d'une variété de cotonnier adaptée aux conditions climatologiques et agrologiques locales: travaux effectués à la Ferme Cotonnière de Ferkésédougou de 1926–1938, rapport par J. Lebeuf, 8 June 1941.
ANRS 1 R 208, Vers. 158, Organisation de la Production Cotonnière en AOF (1942).
ANRS 1 R 208, Vers. 158, Territoire de la Côte d'Ivoire, Circulaire du Service de l'Agriculture a/s culture cotonnière, February 1951, p. 1.
ANRS 1 R 208 d, Vers. 158, Inspecteur Général de l'Agriculture, Note sur la production cotonnière en AOF, January 1952.

ARCHIVES NATIONALES DE LA RÉPUBLIQUE DU MALI (ANRM)

ANRM lN-173, Opérations militaires, Rapports Côte d'Ivoire,1890–99.
ANRM lN-174, Opérations militaires, Rapports Côte d'Ivoire, 1890–99.
ANRM lR-123, Note sur la culture de coton en Côte d Ivoire, 1918.

ARCHIVES NATIONALES DE FRANCE – SECTION D'OUTRE-MER SERIES (ANSOM)

ANSOM AF 1142, Service Géographique des Colonies, Mission Marchand, Le Transnigérien, Le Bandama et le Bagoé, (la 1ère Partie le Bani-Niger a été comprise dans la Carte Binger), Carte Levée et Dresée de 1892–1895 par le Capitaine Marchand, 1/500.000.

ANSOM Aff Pol 3047 bis, Situation de la Culture du Cotonnière et de la Production du Coton au 31 décembre 1917 (Bingerville 1918).

ANSOM Aff Pol 3066, Inspection des Affaires Administratives, Kair Report, 4 March 1931.

ANSOM 2G47–68, Côte d'Ivoire, Service de l'Agriculture, Rapport Annuel – 1947.

ANSOM 14 Mi 1848, 2 G 43 (51), Côte d'Ivoire, Service de l'Agriculture, 'Rapport Annuel 1943.'

ANSOM 14 Mi 1856, 2 G 44 (50), Côte d'Ivoire, Service de l'Agriculture, 'Rapport Annuel 1944'.

ANSOM 14 Mi 1878, 2 G 46, Colonie de la Côte d'Ivoire, Service de l'Agriculture, 'Rapport Annuel 1946.'

ANSOM 14 Mi 2033, 2 G 56 (27), Service de l'Agricuture, 'Rapport Annuel – 1956'.

ANSOM 14 Mi 2696, 2 G 43 (99), Rapport sur la situation politique d'ensemble pendant le mois de mai 1943.

ANSOM Mission 8, Letter from Captain Marchand to the Minister of the Colonies, 20 December 1884.

ANSOM S. G. Soudan I- 9. Letter from Minister of Colonies to the Governor General of AOF on the subject of the French Sudan, instructions for the Lieuteuant Governor de Trentinian, 10 November 1898.

ANSOM S.G. Soudan 4, Letter from M. Ballay, Inspector of the Colonies to the Minister of the Colonies, 13 July 1898.

ANSOM S. G. Soudan 9, Rapport sur la situation politique du Soudan français au 15 fevrier 1896.

ANSOM S. G. Soudan 9, Rapport sur la situation politique du Soudan français au 31 julliet 1896.

ANSOM S.G. Soudan 9, Rapport sur la situation politique du Soudan français au 1 octobre 1896.

ANSOM S.G. Soudan 9, Rapport sur la situation politique du Soudan français au 1 janvier 1897.

ANSOM S.G. Soudan 9, letter from the Governor General of AOF to the Minister of the the Colonies, 31 March 1898.

ANSOM S.G. Soudan 9, Rapport politique au mois de mai, 8 juin 1898.

ANSOM S.G. Soudan 9, letter from the Lieutenant Governor of the French Soudan to the Minister of the Colonies, 28 April 1899, a/s envois de la carte politique du Soudan français.

INTERVIEWS CITED IN THE TEXT

Interviews with Fongnonbêh Tuo, farmer, Katiali, 3 July 1981; 4 April 1982.
Interview with Donisongui Silué, farmer, Katiali, 30 June 1986.
Interview with M. Zahouri, CIDT extension agent, Katiali, 27 Feb 1992.
Interview with Mamadu Dosso, Executive Secretary, URECOS-CI, Korhogo, 2 February 1992.
Interview with Zéa Silué, farmer and secretary of COPAG-CI, Katiali, 14 February 1992.
Interview with Bazoumana Diabaté, farmer, Katiali, 28 February 1992.
Interviews with Katienen'golo Silué, farmer, Katiali, 28 February, 12–13 March 1992.
Interview with Brahima Coulibaly, President of M'Bengué GVC Union, 29 February 1992.
Interview with Gniofolotien Silué, farmer, Katiali, 3 March 1992.
Interview with Adama Koné, Katiali, former-GVC President, 7 March 1992.
Interview with Kaléléna Silué, village chief, Katiali, 12 March 1992.
Interview with M. Cluchier, retired CFDT official, Bouaké, 21 March 1992.
Interview with CIDT monitor, Katiali, 19 August 1995.
Interview with Blais Faustain, CIDT zone chief, Niofouin, 21 August 1995.
Interview with Bernadin Ouattara, retired CFDT extension agent, Korhogo, 23 August 1995.
Interview with Navigué Soro, Korhogo, retired CFDT extension agent, 23 August 1995.
Interview with François Béroud, Director of Rural Development, CFDT, Paris, France, 16 April 1996.

SECONDARY SOURCES

Affou, S. Y., "Renforcement des organisations paysannes et progrès agricole: obstacles ou atouts pour le progrès agricole," in B. Contamin and H. Memel-Fotê (eds.), *Le modèle ivoirien en questions: crises, ajustements, recompositions* (Paris: Karthala and ORSTOM, 1997), pp. 555–71.
Amin, S., *Le développement du capitalisme en Côte d'Ivoire* (Paris: Editions Minuit, 1967).
Amprou, J., L. Gbeli and J. McIntire, "Le coton en Côte d'Ivoire: état des réformes," mimeograph., Banque Mondiale (3 June 1998).
Arnaud, J-C. and J. C. Filleron, "Eléments pour une géographie du peuplement dans le nord-ouest de la Côte d'Ivoire," *Annales de l'Université d'Abidjan, série G (Géographie)*, Vol. 9, 1980.
Aubertin, C., *Histoire et création d'une région "sous-développée" (le Nord Ivoirien)* (Abidjan: ORSTOM, 1980).
Bartram, W., *The travels of William Bartram* (New York: Facsimile Library, 1940).
Bassett, T., "Food, peasantry and the state in northern Côte d'Ivoire, 1893–1982," (Ph.D. thesis, University of California, Berkeley, 1984).
"Fulani herd movements," *Geographical Review* 76 (1986), 233–48.
"Development theory and reality: the World Bank in northern Côte d'Ivoire," *Review of African Political Economy*, 41 (1988), 45–59.

"The political ecology of peasant–herder conflicts in northern Ivory Coast," *Annals of the Association of American Geographers* 78 (1988), 453–72.

"Breaking up the bottlenecks in food-crop and cotton cultivation in northern Côte d'Ivoire," *Africa* 58 (1988), 157–61.

"Migration et féminisation de l'agriculture dans le nord de la Côte d'Ivoire," in C. Meillassoux and B. Schlemmer (eds.), *Déséquilibres démographiques, déséquilibres alimentaires* (Paris: Etudes et Documentations Internationales, 1991), pp. 219–45.

"Land use conflicts in pastoral development in northern Côte d'Ivoire," in T. J. Bassett and D. Crummey (eds.), *Land in African Agrarian Systems* (Madison: University of Wisconsin Press, 1993), pp. 131–54.

"Hired herders and herd management in Fulani pastoralism (northern Côte d'Ivoire)," *Cahiers d'Etudes Africaines* 34 (1994), 147–73.

Bassett, T. and P. Porter, "'From the Best Authorities': the mountains of Kong in the cartography of West Africa," *Journal of African History,* 32, 3 (1991), 367–413.

Bates, R., *Markets and states in tropical Africa: the political basis of agricultural policies* (Berkeley: University of California Press, 1981).

Essays on the political economy of rural Africa (Cambridge: Cambridge University Press, 1983).

Beyond the miracle of the market: the political economy of agrarian development in Kenya (Cambridge: Cambridge University Press, 1989).

Baulin, J., *La politique intérieure d'Houphouet-Boigny* (Paris: Editions Eurafor-Press, 1982).

Bernardet, P., *Association agriculture-élevage en Afrique: les peuls semi-transhumants de Côte d'Ivoire* (Paris: L'Harmattan, 1984).

Bernstein, H., "African peasantries: a theoretical framework," *Journal of Peasant Studies* 6 (1979), 420–43.

"Notes on capital and peasantry," *Review of African Political Economy*, 10 (1978), 60–73.

Béroud, F., "Coton et Environnement: des idées reçues dont il faut se méfier," *Afrique Agriculture* 271 (1999), 59–61.

Berry, S., "The food crisis and agrarian change in Africa: a review essay," *African Studies Review* 27, 2 (1984), 59–112.

No condition is permanent: the social dynamics of agrarian change in sub-Saharan Africa (Madison: University of Wisconsin Press, 1993).

Beusekom, M. van, "Contested development: African farmers, colonial officials, and agricultural practices at the Office du Niger, 1920–60," paper presented at the annual meeting of the African Studies Association, Columbus, Ohio, 13–16 Nov. 1997.

Biggs, S. D. and E. J. Clay, "Generation and diffusion of agricultural technology: a review of theories and experiences," *World Employment Programme Research Working Paper*, 122 (Geneva: ILO, 1983).

Bigot, Y., "Analyse techno-économique des systèmes de production coton-céréale dans le Nord-Ouest de la Côte d'Ivoire" (Bouaké: Institut de Savanes, 1979).

"La culture attelée et ses limites dans l'évolution des systèmes de production en zone de savane en Côte d'Ivoire," *Cahiers du CIRES* 30 (1981), 9–30.

Bingen, J., "Agricultural development policy and grassroots democracy in Mali: the emergence of Mali's farmer movement," *African Rural and Urban Studies* 1 (1994), 57–72.

Binger, Le Capitaine, "Du Niger au Golfe de Guinée," *Le Tour du Monde* LXI (1891), I: 1–128; II: 33–144.

Binger, L. G., *Du Niger au Golfe de Guinée, par le pays de King et le Mossi (1887–1889)*, 2 vols. (Paris: Hachette, 1892).

Bratton, M. and R. J. Bingen, "Farmer organization and agricultural policy in Africa – Introduction," *African Rural and Urban Studies* 1 (1994), 7–30.

Brubaker, C. L., F. M. Bourland and J. F. Wendel, "The origin and domestication of cotton," in C. W. Smith and J. T. Cothren (eds.), *Cotton: origin, history, technology, and production* (New York: John Wiley and Sons, 1999), pp. 3–31.

Brush, S. and W. Turner, *Comparative farming systems* (New York: The Guilford Press, 1987).

Burmeister, L., "The South Korean green revolution: induced or directed innovation?" *Economic Development and Cultural Change* 35 (1987), 766–90.

"Induced innovation and agricultural research in South Korea," in B. Koppel (ed.), *Induced innovation theory and international agricultural development: a reassessment* (Baltimore: Johns Hopkins U. Press, 1995), pp. 39–55.

Byerlee, D. and C. Eicher, *Africa's emerging maize revolution* (Boulder: Lynne Rienner, 1997).

Caillié, R., *Journal d'un voyage à Temboctou et à Jenné dans l'Afrique centrale* (Paris: l'Imprimerie Royale, 1830).

Campbell, B.,"Inside the miracle: cotton in the Ivory Coast," in J. Barker (ed.) *The Politics of Tropical Agriculture* (Beverly Hills, CA: Sage, 1984), pp. 143–71.

"The fiscal crisis of the state: the case of the Ivory Coast," in H. Bernstein and B. Campbell (eds.), *Contradictions of accumulation in Africa: studies in economy and society* (Beverly Hills, CA: Sage, 1985), pp. 267–310.

"Le modèle ivoirien de développement à l'épreuve de la crise," in B. Contamin and H. Memel-Fotê (eds.), *Le modèle ivoirien en questions: crises, ajustements, recompositions* (Paris: Karthala and ORSTOM, 1997), pp. 37–60.

Carbon Ferrière, J. de, "Le développement de la CFDT," in F. Bocchino, *et al.* (eds.) *Cinquante ans d'action cotonnière au service du développement*, special issue of *Coton et Développement* (September 1999), pp. 37–41.

Carney, J. and M. Watts, "Manufacturing dissent: work, gender, and the politics of meaning in peasant society," *Africa* 60 (1990), 207–41.

Chaleard, J-L., *Temps de villes, temps des vivres: l'essor du vivrier marchand en Côte d'Ivoire* (Paris: Karthala, 1996).

Chauveau, J.-P., "L'étude des dynamiques agraires et la problématique de l'innovation," in J.-P. Chauveau, M.-C. Cormier-Salem, and E. Mollard (eds.), *L'innovation en agriculture: questions de méthodes et terrains d'observation* (Paris: IRD, 1999), 9–31.

Chevalier, A., "L'avenir de la culture du cotonnier au Soudan Français," *Bulletin de la Société Nationale d'Acclimatation de France* (1901) 225–44.

"La question de la culture des cotonniers en Afrique tropicale," *Comptes Rendus des Seances de l'Academie des Sciences* (1904), 1–3.

CIDT, *Rapport Annuel* (Bouaké: CIDT, various years).

CIDT, *Annuaire Signalétique: CIDT 1995–1996* (Bouaké: CIDT, 1996).

Cleaver, K. and G. Schreiber, *Reversing the spiral: the population, agriculture, and environment nexus in sub-Saharan Africa* (Washington, DC: The World Bank, 1994).

Clerin, R., "La production cotonnière au Soudan Français;" in E. Guernier (ed.), *L'Encyclopédie Coloniale et Maritime*, Vol. 2, *Afrique Occidentale Française* (Paris: Encyclopédie Coloniale et Maritime, 1949).

Clifford, J., "Power and dialogue in ethnography: Marcel Griaule's initiation," in G. Stocking (ed.), *Observers Observed: essays on ethnographic fieldwork* (Madison: University of Wisconsin Press, 1983).

Colnard, C., "Pratiques paysannes et utilisation des intrants en culture cotonnière au Benin" (Mémoire de Fin d'Etudes, Ecole Nationale Supérieure d'Horticulture de Versailles, 1995).

Contamin, B. and Y. Fauré, *La bataille des entreprises publiques en Côte d'Ivoire: histoire d'un ajustement interne* (Paris: Karthala, 1990).

Contamin, B., "Entreprises publiques et désengagement de l'Etat en Côte d'Ivoire: à la recherche des privatisations," in B. Contamin and H. Memel-Fotê (eds.), *Le modèle ivoirien en questions: Crises, ajustements, recompositions* (Paris: Karthala and ORSTOM, 1997), pp. 89–107.

Contamin, B. and H. Memel-Fotê (eds.), *Le Modèle ivoirien en questions: crises, ajustements, recompositions* (Paris: Karthala and ORSTOM, 1997).

Cooper, F., *Decolonization and African society: the labor question in French and British Africa* (Cambridge: Cambridge University Press, 1996).

Coulibaly, Sinali, *Le paysan sénoufo* (Abidjan: Nouvelles Editions Africaines, 1978).

"Etat, société et développement: le cas Sénoufo dans le nord Ivoirien" (Thèse de doctorat d'état, Université de Paris X, 1990).

Coulibaly, Samba, "Le rôle de la CIDT est primordial: entretien avec M. Samba Coulibaly, directeur général de la CIDT," *Coton et Développement* 21 (1997), 5–11.

Crush, J. (ed.), *Power of development* (London: Routledge, 1995).

Cuzon, J-R., "Réalités des pratiques paysannes en matière d'utilisation des intrants sur le coton au Sénégal" (Mémoire Diplôme d'Agronomie Tropicale, Ecole Nationale Supérieure des Sciences Agronomiques Appliquées, 1993) .

David, P., "La carte postale ivoirienne de 1900 à 1960: un bilan iconographique et culturel provisoire," *Notes Africaines* 174 (1982), 29–39.

De Janvry, A., M. Fafchamps and E. Sadoulet, "Transaction costs, public choice, and induced technological innovations," in B. Koppel (ed.), *Induced innovation theory and international agricultural development: a reassessment* (Baltimore, Johns Hopkins University Press, 1995), pp. 151–68.

Delafosse, M., "Le Peuple Siéna ou Sénoufo," *Revue des études ethnographiques et sociologiques*, 2 vols. (1908–09), Vol. 1: 16–32, 79–92,151–8, 242–75, 448–57; Vol. 2: 1–21.

Demery, L., "Côte d'Ivoire: fettered adjustment," in I. Husain and R. Faruqee (eds.), *Adjustment in Africa: lessons from country case studies* (Washington, DC: The World Bank, 1994), pp. 72–152.

Demont, M. and P. Jouve," Evolution d'agroécosystèmes villageois dans la région de Korhogo (Nord Cote d'Ivoire): révision des débats 'Boserup versus

Malthus' et compétition versus complémentarité," paper presented at the conference "Dynamiques agraires et construction sociale du territoire," Montpellier, 26–28 April 1999.

Deveze, J-C., "Les zones cotonnières entre développement, ajustement et dévaluation. Réflexions sur le rôle du coton en Afrique francophone de l'ouest et du centre," *Caisse Française de Développement, Notes et Etudes*, 53 (1994).

Dibi, P., "Regional climate at the Ivorian-Burkina borderlands," paper presented at the symposium, "African savannas: new perspectives on environmental and social change," Urbana, Illinois, 3 April, 1998.

Dinham, B. and C. Hines, *Agribusiness in Africa* (London: Earth Resources Research, 1983).

Dupont, S., "La guerre du coton noir: Américains et Français s'opposent en Afrique," *Les Echos* (16–17 February 1996), 50–51.

"Le Cameroun ou les dangers d'une libéralisation incontrôlée," *Les Echos* (16–17 February 1996), 51.

Eponou, T., "Farm level analysis of rice production systems in northwestern Ivory Coast", (Ph.D. thesis, Michigan State University, 1983).

Escobar, A., *Encountering development: the making and unmaking of the Third World* (Princeton: Princeton University Press, 1995).

"Imagining a post-development era," in J. Crush (ed.), *Power of development* (London: Routledge, 1995), pp. 211–27.

Fall, B., *Le travail forcé en Afrique Occidentale française, 1900–1946* (Paris: Karthala, 1993).

Fargues, P., "La dynamique démographique des producteurs de vivriers," in *Les cultures vivrières: élément stratégique du développement agricole Ivoirien* (Abidjan: CIRES, 1983), Vol. 1, pp. 83–104.

Fauré, Y-A., "Le complexe politico-économique," in Y-A Fauré and J-F Médard (eds.), *Etat et bourgeoisie en Côte d'Ivoire* (Paris: Karthala, 1982), pp. 21–60.

Fauré, Y-A. and J-F. Médard, *Etat et bourgeoisie en Côte d'Ivoire* (Paris: Karthala, 1982).

Ferguson, J., *The Anti-Politics machine: "development," depoliticization, and bureaucratic power in Lesotho* (Minneapolis: University of Minnesota Press, 1990).

"The anti-politics machine: development and bureaucratic power in Lesotho," *The Ecologist* 24 (1994), 176–81.

Fichet, M., "Côte d'Ivoire: Derrière le concept de privatisation, le démantèlement d'une filière," *Coton et Développement* 22 (1997), 2–3. Reprinted from the Côte d'Ivoire newspaper, *La Voie*, 21 avril 1997.

Fok, M., *Le développement du coton au Mali par analyse des contradictions: les acteurs et les crises de 1895 à 1993* (CIRAD, Unité de Recherche Economie des Filières No. 8, 1994).

Gastellu, J-M, *Riches paysans de Côte-d'Ivoire* (Paris: L'Harmattan, 1989).

Gastellu, J-M and S. Yaffou Yapi, "Un mythe à décomposer: la bourgeoisie de planteurs," in Y-A Fauré and J-F Médard (eds.), *Etat et bourgeoisie en Côte d'Ivoire* (Paris: Karthala, 1982), pp. 149–79.

Gatelet, Lieutenant, *Histoire de la Conquête du Soudan Français (1878–1899)* (Paris: Berger-Levrault and Compagnie, 1901).

Gbagbo, L., *Côte d'Ivoire: économie et société à la veille de l'indépendance (1940–1960)* (Paris: L'Harmattan, 1982).

Geay, F. and C. Konomou, *Pratiques paysannes en culture cotonnière*, Projet de développement rural de la Haute-Guinée, Suivi-Evaluation (Conakry: PDRHG, 1993).

GERDAT, *Intensification de l'agrosystème en cultures assolées dans le Nord de la Côte d'Ivoire* (Bouaké: GERDAT, 1979).

Giddens, A., *Central problems in social theory: action, structure and contradiction in social analysis* (Berkeley: University of California Press, 1979).

Giraudy, F., *Annuaire statistique 94/95: résultats de l'enquête agricole permanente* (Sikasso: CMDT, 1995).

Glaze, A., *Life and death in a Senufo village* (Bloomington: Indiana University Press, 1981).

Grabowski, R., "Induced innovation: a critical perspective," in B. Koppel (ed.), *Induced innovation theory and international agricultural development: a reassessment* (Baltimore: Johns Hopkins University Press, 1995), pp. 73–92.

Gray, L., *Land degradation in southwestern Burkina Faso: the environmental effects of demographic and agricultural change* (Ph.D. thesis, University of Illinois, Urbana-Champaign, 1997).

Grigg, D., *The dynamics of agricultural change: the historical experience* (New York: Saint Martin's Press, 1982).

Groupe de Travail Coopération Française, *Le coton en Afrique de l'Ouest et du Centre: situation et perspectives* (Paris: Ministère de la Coopération Française, 1991).

Handloff, R., "The Dyula of Gyaman: A study of politics and trade in the 19th century" (Ph.D. thesis, Northwestern University, 1982).

Hecht, R., "The Ivory Coast miracle: what benefits for peasant producers?" *Journal of Modern African Studies*, 21 (1983), 25–53.

Herbel, D., *La compétitivité du coton ivoirien*, Document de travail (Montpellier: CIRAD, 1995).

Hermann, E., *Analysis of selected agricultural parastatals in the Ivory Coast* (Abidjan: REDSO/WA, 1981).

Hoffman, O., *Pratiques pastorales et dynamique du couvert végétale en pays Lobi (Nord est de la Côte d'Ivoire)*, (Paris: ORSTOM, Collection Travaux et Documents 189, 1985).

Hogendorn, J., "The cotton campaign in northern Nigeria, 1902–1914: an example of public/private planning failure in agriculture," in A. Isaacman and R. Roberts (eds.), *Cotton, colonialism, and social history in sub-Saharan Africa* (Portsmouth, NH: Heinemann, 1995), pp. 50–70.

Holas, H., *Les Sénoufo (y compris les Minianka)* (Paris: Presses Universitaires de France, 1957).

Hutchinson, J. B., R. A. Silow and S. G. Stephens, *The Evolution of Gossypium* (London: Oxford University Press, 1947).

Hutchinson, J., *The Application of genetics to cotton improvement* (Cambridge: Cambridge University Press, 1959).

IRCT, *La culture cotonnière dans le nord de la Côte d'Ivoire* (Bouaké: IRCT, 1976).

Isaacman, A., *Cotton is the mother of poverty: peasants, work, and rural struggle in colonial Mozambique, 1938–1961* (Portsmouth, NH: Heinemann, 1996).

Isaacman, A. and R. Roberts (eds.), *Cotton, colonialism, and social history in sub-Saharan Africa* (Portsmouth, NH: Heinemann, 1995).

Kanya-Forstner, A.S., *Conquest of the western Soudan: a study in French military imperialism* (Cambridge: Cambridge University Press, 1969).

Keyes Aderoike, C. and J. Vansina (eds.), *In pursuit of history: fieldwork in Africa* (Portsmouth, NH: Heinemann, 1995).

Kientz, A., "Approches de parentés sénoufo," *J. des Africanistes* 49 (1979), 9–28.

Pour une motorisation paysanne: mécanisation et motorisation des exploitations paysannes, Côte d'Ivoire, région nord (Abidjan: Ministère du Développement Rural, 1985).

Développement agro-pastoral et lutte anti-tsé-tsé, Côte d'Ivoire (GTZ Report PN 87.2539.2–01.100, 1991).

Koppel, B., "Why a reassessment?" in B. Koppel (ed.), *Induced innovation theory and international agricultural development: a reassessment* (Baltimore: Johns Hopkins University Press, 1995), pp. 3–21.

Koppel, B. (ed.), *Induced innovation theory and international agricultural development: a reassessment* (Baltimore: Johns Hopkins University Press, 1995).

Kowal, J. M. and K. B. Adeoye, "An assessment of aridity and the severity of the 1972 drought in northern Nigeria and neighboring countries," *Savanna* 2 (1973), 145–58.

Kowal, J. M. and A. H. Kassam, *Agricultural ecology of savanna: a study of West Africa* (Oxford: Clarendon Press, 1978).

Labouret, H. *Paysans d'Afrique occidentale* (Paris: Gallimard, 1941).

Launay, R., *Traders without trade* (Cambridge: Cambridge University Press, 1982).

"Stereotypic vision: the 'moral character' of the Senufo in colonial and postcolonial discourse," *Cahiers d'Etudes Africaines*, 49 (1999), 271–92.

Lebeuf, J., "Le coton en Côte d'Ivoire et en Haute-Volta," in E. Guernier (ed.), *L'Encyclopédie Coloniale et Maritime*, Vol. 2, *Afrique Occidentale Française* (Paris: Encyclopédie Coloniale et Maritime, 1949).

Lee, E., "Export-led rural development: the Ivory Coast," *Development and Change* 11 (1980), 607–42.

Lele, U., N. van de Walle, and M. Gbetibouo, "Cotton in Africa: an analysis of differences in performance," *MADIA Discussion Paper 7* (Washington, DC: The World Bank, 1989).

Lele, U. and S. Stone, *Population pressure, the environment, and agricultural intensification: variations on the Boserup hypothesis*, MADIA Discussion Paper 4 (Washington, DC: The World Bank, 1989).

Lendres, P., *Pratiques paysannes et utilisation des intrants en culture cotonnière au Burkina Faso*, Mémoire de Fin d'Etudes, Centre National d'Etudes Agronomiques des Régions Chaudes (1993).

Leroide, H., *Situation de la culture du cotonnier et de la production du coton au 31 décembre 1917* (Bingerville: Colonie de la Côte d'Ivoire, 1918).

Le Roy, X., *L'introduction des cultures de rapport dans l'agriculture vivrière sénoufo: le cas de Karakpo* (Abidjan: ORSTOM, 1980).

"Où la culture cotonnière rénove l'économie paysanne (Côte d'Ivoire)," in J.-P. Chauveau, M.-C. Cormier-Salem, and E. Mollard (eds.), *L'innovation en agriculture: questions de méthodes et terrains d'observation* (Paris: IRD, 1999), pp. 199–212.

Likaka, O., *Rural society and cotton in colonial Zaire* (Madison: University of Wisconsin Press, 1997).

Little, P. D., "The link between local participation and improved conservation: a review of the issues and experiences," in D. Western, *et al.* (eds), *Natural connections: perspectives in community-based conservation* (Washington, DC: Island Press, 1994), pp. 347–72.

Little, P. D. and M.J. Watts, *Living under contract: contract farming and agrarian transformation in sub-Saharan Africa* (Madison: University of Wisconsin Press, 1994).

Londres, A., *Terre d'Ebène* (Paris: Arléa, 1998 [1928]).

Mahop, F. and E. Van Ranst, "Côut de l'épuisement du sol en zone cotonnière camerounaise: impact sur l'environnement," *Tropicultura* 15 (1997), 203–8.

Maier, D., "Persistence of pre-colonial patterns of production: cotton in German Togoland," in A. Isaacman and R. Roberts (eds.), *Cotton, colonialism, and social history in sub-Saharan Africa* (Portsmouth, NH: Heinemann, 1995), pp. 71–95.

Manning, P., *Francophone Sub-Saharan Africa, 1880–1985* (Cambridge: Cambridge University Press, 1988).

Marcussen, H. and J. Torp, *Internationalization of capital: prospects for the Third World* (London: Zed Press, 1982).

Martin, F., "Compagnie française pour le développement des textiles (CFDT)," *Revue Internationale des Produits Coloniaux* 28 (1952), 115–20.

Mitchell, T., "The object of development," in J. Crush (ed.), *Power of development* (London: Routledge, 1995), pp. 129–57.

Marty, P., *Etudes sur l'Islam en Côte d'Ivoire* (Paris: E. Leroux, 1922).

Meillassoux, C., *Anthropologie économique des Gouros de Côte d'Ivoire* (Paris: Mouton, 1964).

"From reproduction to production," *Economy and Society*, 1 (1972), 93–105.

"The economic basis of demographic reproduction: from the domestic mode of production to wage economy," *Journal of Peasant Studies* 11 (1983), 50–61.

Méniaud, J., *Les Pionniers du Soudan*, 2 vols. (Paris: Société des Publications Modernes, 1931).

Michelis, J., "Pourqoui la CFDT?" in F. Bocchino, *et al.* (eds.), *Cinquante ans d'action cotonnière au service du développement*, special issue of *Coton et Développement* (September 1999), 32–6.

Ministère de la Coopération Française, *Le coton en Afrique de l'Ouest et du centre: situation et perspectives* (première édition) (Paris: La Documentation française, 1987).

Mitja, D., *Influence de la culture itinérante sur la végétation d'une savane humide de Côte d'Ivoire*, Collection Etudes et Thèses (Paris: ORSTOM, 1992).

Mitchell, T., "The object of development," in J. Crush (ed.), *Power of development* (London: Routledge, 1995), pp. 129–57.

Monnoyer, C., "Le problème de la culture du cotonnier en Afrique occidentale française," *Revue de Botanique Appliquée et d'Agriculture Coloniale* 2 (1922), 613.

Monteil, P.L., *De St Louis à Tripoli par le Lac Tchad* (Paris: 1984).

Moore, D., "Marxism, culture, and political ecology," in R. Peet and M. Watts (eds.) *Liberation ecologies* (London: Routledge, 1996), pp. 125–47.

Munro, J. M., *Cotton* (Essex: Longman Scientific and Technical, 1987).

Norman, D., "Farming systems and problems of improving them," in J. Kowal and

A.H. Kassam (eds.), *Agricultural ecology of savanna: a study of West Africa* (Oxford: Clarendon Press, 1978), pp. 318–40.

Nyong'o, P. Anyang, "Liberal models of capitalist development in Africa: Ivory Coast," *Africa Development* 3 (1978), 5–20.

Olivier de Sardan, J-P, *Anthropologie et développement: essai en socio-anthropologie du changement social* (Paris: Karthala, 1996).

Ouattara, T., "Les Tiembara de Korhogo, des origines à Péléforo Gbon Coulibaly (1962)" (Thèse de Doctorat de 3ème cycle d'Histoire, Université de Paris, 1977).

Perron, M., "Situation économique et agricole du pays Sénoufo: notre rôle", *Bulletin de l'agence générale des colonies* (1933).

Person, Y., *Samori: Une révolution Dyula,* Mémoires de l'Institut Fondamental d'Afrique Noire, No. 80, 3 vols. (Dakar: IFAN, 1968, 1970, 1975).

Cartes historiques de l'Afrique manding (fin du 19e siecle). Samori: Une révolution Dyula (Paris: Centre de Recherches Africaines, 1990).

Peltre-Wurtz, J. and B. Steck, *Les Charrues de la Bagoé* (Paris: ORSTOM, 1991).

Pingali, P. L., Y. Bigot and H. Binswanger, *Agricultural mechanization and the evolution of farming systems in sub-Saharan Africa* (Baltimore: Johns Hopkins University Press, 1987).

Prentice, A.N., *Cotton: with special reference to Africa* (London: Longman, 1972).

Pursell, G., "Cotton policies in Francophone Africa," (manuscript, World Bank, 21 May 1998).

Rabinow, P., *Reflections on fieldwork in Morocco* (Berkeley: University of California Press, 1977).

Raingead, J. and C. Romuald-Robert, *Rapport annuel pour la campagne 1955–1956*, (Bouaké: IRCT, 1956).

Rapley, J., *Ivorian capitalism: African entrepreneurs in Côte d'Ivoire* (Boulder, CO: Lynne Rienner, 1993).

République de Côte d'Ivoire (RCI), Min. de l'Agriculture et des Eaux et Forêts, Direction de la Planification et de la Programmation Agricoles et Forestières, *Bilan Diagnostic du Secteur Agricole en Côte d'Ivoire 1: Filière Coton*, Document de Travail, Abidjan, Juin 1990.

République de Côte d'Ivoire (RCI), Ministère de l'Agriculture et des Eaux et Forêts, Ministère de l'Economie et des Finances, *Prêt d'ajustement sectoriel de l'agriculture: Etude des conditions d'équilibre de la filière coton; Preparation d'un contrat-plan entre l'état et la CIDT. Provisoire*, Novembre 1990 (Abidjan: Direction et Controle des Grands Travaux).

Richards, P., *Indigenous agricultural revolution: ecology and food production in West Africa* (London and Boulder, CO: Hutchinson and Westview Press, 1985).

Coping with hunger: hazard and experiment in an African rice-farming system (London: Allen and Unwin, 1986).

Roberts, R., *Two worlds of cotton: colonialism and the regional economy in the French Soudan, 1800–1946* (Palo Alto: Stanford University Press, 1997).

Roe, E., "Development narratives, or making the best of blueprint development," *World Development* 19 (1991), 287–300.

Romuald-Robert, C., *Amélioration de la production cotonnière en Côte d'Ivoire* (Bouaké: IRCT, 1962).

Rondeau, C., "La société Senoufo du sud Mali (1870–1950): de la "tradition" à la dépendance", (Thèse de doctorat de 3ème cycle, Université de Paris, France, 1980).

Roupsard, M., "Nord-Cameroun: ouverture et développement d'une région enclavée" (Thèse de doctorat, Université de Paris, 1987).

Ruf, F., *Booms et crises du cacao: les vertiges de l'or brun* (Paris: Karthala, 1995).

Ruttan, V., "Induced innovation and agricultural development," in G.K. Douglass (ed.), *Agricultural sustainability in a changing world order* (Boulder: Westview Press, 1984), pp. 107–34.

Ruttan, V. and Y. Hayami, "Induced innovation theory and agricultural development: a personal account," in B. Koppel (ed.), *Induced innovation theory and international agricultural development: a reassessment* (Baltimore: Johns Hopkins University Press, 1995), pp. 22–36.

Sawadogo, A., *L'agriculture en Côte d'Ivoire* (Paris: Presses Universitaires de France, 1977).

Scott, J., *Weapons of the weak: everyday forms of resistance* (New Haven: Yale University Press, 1985).

"Resistance without protest and without organization: peasant opposition to the Islamic *zakat* and the Christian tithe," *Comparative Studies in Society and History* 29 (1987), 417–24

SEDES, *Région de Korhogo, Etude de Développement Socio-Economique*: Vol. 1, *Rapport Démographique;* Vol. 2, *Rapport Sociologique*; Vol. 3 *Rapport Agricole*, Vol. 4, *Rapport Pedologique*; Vol. 5, *Etudes des budgets familiaux* (Paris: SEDES, 1965).

SEDES, *Etude Régionale de Bouaké 1962–64*, Vol. 2, *L'Economie* (Paris: SEDES, 1965).

Sement, G., "Etude des effets secondaires de la fertilisation minérale sur le sol dans des systèmes culturaux à base de coton en Côte d'Ivoire: premier résultats en matière de correction," *Coton et fibres tropicales* 35 (1990), 229–48.

Sigrist, J-C., "Pratiques paysannes et utilisation des intrants en culture cotonnière au nord-Cameroun", (Mémoire de Fin d'Etudes, Institut Supérieur Technique d'Outre-Mer, Cergy-Pontoise, France, 1992).

Spindel, C., *In the shadow of the sacred grove* (New York: Vintage, 1989).

Staatz, J. and C. Eicher, "Agricultural development ideas in historical perspective," in *Agricultural development in the Third World* (Baltimore: Johns Hopkins University Press, 1990), pp. 3–37.

Stephens, S. G., "The origins of Sea Island cotton," *Agricultural History* 50 (1976), 391–9.

Suret-Canale, J., *French colonialism in tropical Africa, 1900–1945* (New York: Pica Press, 1971).

Tautain, L. "Le Dioula-dougou et le Sénéfo," *Revue d'Ethnographie* VI (1887), 395–8.

Tosh, J., "Lango agriculture during the early colonial period: land and labour in a cash-crop economy," *Journal of African History* 19 (1978), 415–39.

Tuinder, B. den, *Ivory Coast: the challenge of success* (Baltimore: Johns Hopkins University Press, 1978).

US Department of Agriculture (USDA), Foreign Agricultural Service, *Cotton Annual – 1999*. Gain Report IV9011 (1999).

US Department of Agriculture (USDA), Foreign Agricultural Service, "Côte d'Ivoire, cotton producer price," Gain Report IV8022, 11 December 1998.

Vaissayre, M. "La résistance aux pesticides des ravageurs: une menace pour la culture cotonnière en Afrique?" *Afrique Agriculture* 260 (1998), 65–6.

Van de Walle, N., "The decline of the franc zone: monetary politics in francophone Africa," *African Affairs* 90 (1991), 383–406.

Watts, M., *Silent violence: food, famine and peasantry in northern Nigeria* (Berkeley: University of California Press, 1983).

"Development I: power, knowledge, discursive practice," *Progress in Human Geography* 17(1993), 257–72.

"Development II: the privatization of everything?" *Progress in Human Geography* 18 (1994), 371–84.

Watts, M. and T. Bassett, "Crisis and change in African agriculture: a comparative study of the Ivory Coast and Nigeria," *African Studies Review* 28 (1985), 3–27.

Weiner, D., S. Moyo, B. Munslow, and P. O'Keefe, "Land use and agricultural productivity in Zimbabwe," *Journal of Modern African Studies* 23(1985), 251–85.

Weiskel, T., *French colonial rule and the Baoule peoples: resistance and collaboration, 1889–1911* (Oxford: Clarendon Press, 1980).

Wendel J. F., "Cotton: Gossypium (Malvaceae)," in J. Smartt and N.W. Simmonds (eds.), *Evolution of crop plants* (London: Longman, 1995), pp. 358–66.

Widner, J., "Political reform in anglophone and francophone African countries," in J. Widner (ed.), *Economic change and political liberalization in sub-Saharan Africa* (Baltimore: Johns Hopkins University Press, 1994), pp. 49–79.

"The discovery of 'politics': smallholder reactions to the cocoa crisis of 1988–90 in Côte d'Ivoire," in T. M. Callaghy and J. Ravenhill (eds.), *Hemmed in: responses to Africa's economic decline* (New York: Columbia University Press, 1993), pp. 279–331.

"The origins of agricultural policy in Ivory Coast, 1960–86," *Journal of Development Studies* 29 (1993), 25–59.

World Bank, International Finance Corporation, *Report and Recommendation of the President of the Board of Directors on a Proposed Investment in Etablissements R. Gonfreville (ERG), S.A. Ivory Coast* (Report No. IFC/P-249) (Washington, DC: The World Bank, 1976).

World Bank, *Accelerated development in sub-Saharan Africa* (Washington, DC: The World Bank, 1981).

World Bank, *Cotton development programs in Burkina, Côte d'Ivoire, and Togo* Operations Evaluation Department (Washington, DC: The World Bank, 1988).

World Bank, *Côte d'Ivoire: revue du secteur agricole*, document du travail préparé pour l'atelier de la revue du secteur agricole (7–10 November 1994).

Index

Note page numbers in *italics* refer to figures; page numbers in **boldface** refer to plates.

Other books in the series

64 *Bankole-Bright and Politics in Colonial Sierra Leone: The Passing of the 'Krio Era', 1919–1958* Akintola Wyse
65 *Contemporary West African States* Donal Cruise O'Brien, John Dunn and Richard Rathbone
66 *The Oromo of Ethiopia: A History, 1570–1860* Mohammed Hassen
67 *Slavery and African Life: Occidental, Oriental and African Slave Trades* Patrick Manning
68 *Abraham Esau's War: A Black South African War in the Cape, 1899–1902* Bill Nasson
69 *The Politics of Harmony: Land Dispute Strategies in Swaziland* Laurel Rose
70 *Zimbabwe's Guerrilla War: Peasant Voices* Norma Kriger
71 *Ethiopia: Power and Protest: Peasant Revolts in the Twentieth Century* Gebru Tareke
72 *White Supremacy and Black Resistance in Pre-Industrial South Africa: The Making of the Colonial Order in the Eastern Cape, 1770–1865* Clifton C. Crais
73 *The Elusive Granary: Herder, Farmer, and State in Northern Kenya* Peter D. Little
74 *The Kanyok of Zaire: An Institutional and Ideological History to 1895* John C. Yoder
75 *Pragmatism in the Age of Jihad: The Precolonial State of Bundu* Michael A. Gomez
76 *Slow Death for Slavery: The Course of Abolition in Northern Nigeria, 1897–1936* Paul E. Lovejoy and Jan S. Hogendorn
77 *West African Slavery and Atlantic Commerce: The Senegal River Valley, 1700–1860* James Searing
78 *A South African Kingdom: The Pursuit of Security in Nineteenth-Century Lesotho* Elizabeth A. Eldredge
79 *State and Society in Pre-Colonial Asante* T. C. McCaskie
80 *Islamic Society and State Power in Senegal: Disciples and Citizens in Fatick* Leonardo A. Villalon
81 *Ethnic Pride and Racial Prejudice in Victorian Cape Town: Group Identity and Social Practice* Vivian Bickford-Smith
82 *The Eritrean Struggle for Independence: Domination, Resistance and Nationalism, 1933–1941* Ruth Iyob
83 *Corruption and State Politics in Sierra Leone* William Reno
84 *The Culture of Politics in Modern Kenya* Angelique Haugerud
85 *Africans: The History of a Continent* John Iliffe
86 *From Slave Trade to 'Legitimate' Commerce* Robin Law

87 *Leisure and Society in Colonial Brazzaville* Phyllis M. Martin
88 *Kingship and State: The Buganda Dynasty* Christopher Wrigley
89 *Decolonization and African Life: The Labour Question in French and British Africa* Frederick Cooper
90 *Misreading the Landscape: Society and Ecology in an African Forest Savannah Mosaic* James Fairhead and Melissa Leach
91 *Peasant Revolution in Ethiopia: The Tigray People's Liberation Front, 1975–1991* John Young
92 *Senegambia and the Atlantic Slave Trade* Boubacar Barry
93 *Commerce and Economic Change in West Africa: The Oil Trade in the Nineteenth Century* Martin Lynn
94 *Slavery and French Colonial Rule in West Africa: Senegal, Guinea and Mali* Martin Klein
95 *East African Doctors: A History of the Modern Profession* John Iliffe
96 *Middleman of the Cameroons Rivers: The Duala and their Hinterland* Ralph A. Austen and Jonathan Derrick
97 *Masters and Servants of the Cape Eastern Frontier, 1760–1803* Susan Newton-King
98 *Status and Respectability in the Cape Colony, 1750–1870* Robert Ross
99 *Slaves, Freedmen and Indentured Laborers in Colonial Mauritius* Richard B. Allen
100 *Transformations in Slavery: A History of Slavery in Africa* Paul E. Lovejoy

Printed in the United States
By Bookmasters